EMPIRE IN PINE

The Story of Lumbering in Wisconsin

1830-1900

AWARDED THE DAVID CLARK EVEREST PRIZE

IN WISCONSIN ECONOMIC HISTORY

EMPIRE IN PINE

THE STORY OF LUMBERING IN WISCONSIN 1830–1900

by Robert F. Fries

Wm Caxton Ltd
Sister Bay, Wisconsin
Evanston, Illinois

This edition published by Wm Caxton Ltd, Box 709, Sister Bay, WI 54234

First published 1951 by the State Historical Society of Wisconsin, Madison, Wisconsin.

Printed in the United States of America

10 9 8 7 6 5 4 3 2 1

Library of Congress Cataloging in Publication Data

Fries, Robert F., 1911-
Empire in pine: the story of lumbering in Wisconsin 1830–1900/
by Robert F. Fries. – Rev. ed.
p. cm.
Bibliography: p.
Includes index.
ISBN 0-940473-10-0 : $25.00. – ISBN 0-940473-11-9 (pbk.) : $14.95
1. Lumber trade--Wisconsin--History--19th century. 2. Logging--Wisconsin--History--19th century. I. Title.
HD9757.W5F7 1989
338.4'7674'09775 – dc20 89-15724
CIP

ISBN# 0-940473-10-0 hardcover
0-940473-11-9 paperback

To

FRANCES CLEMENTS FRIES

Preface

TO write the history of an industry whose roots lie deep in the past while its influence continues to extend into an indefinite future is a somewhat rash undertaking. The lines of such an industry's development interweave in a weirdly complex pattern of cause and effect. In the process of growth it comes to have an astonishingly large number of facets, each of which reflects an entirely different aspect of the industry. The history of lumbering must look very different to the student of logging techniques from what it does to the labor economist or the conservationist.

This book analyzes, from various points of view, the development of the lumber industry in Wisconsin during the exploitive stage. The arrangement is topical. The first eight chapters are devoted to the economic and geographic conditions out of which the industry grew and the changing techniques and organizations which constituted its *modus operandi.* The rest of the book outlines the industry's internal problems and conflicts, the political, economic, and social forces shaping its development, and its growing influence upon the life of the state and nation.

To facilitate analysis and generalization I have omitted—even at the risk of oversimplifying—all details that seem unnecessary to support or illustrate the main points. Accordingly many of the favorite names and anecdotes of the romanticist and antiquarian are missing. Comparatively few lumber companies and lumber barons are mentioned by name. On the other hand, I have included accounts of certain events and developments outside Wisconsin, notably federal government policy, which point up the fact that the industry of the state was part of a larger economy. Statistics, which cannot be entirely avoided in economic history, have been used as sparingly as possible and have been incorporated into the text rather than set up as formal tabulations.

Lumbering in Wisconsin has not of course been confined to the years covered by this study; it has carried over into the present time. But about 1910, with the implementation of conservation sentiment, it entered a new phase. The exciting period of unimpeded exploitation had ended, and what may be called controlled logging was being introduced. No exact date can be assigned to this change; in some parts of the state and in some departments of the industry it came earlier than in others. In this book the emphasis is upon the character of the industry rather than upon the time when a given development was initiated or terminated.

I am indebted to many people and several institutions for countless courtesies. I wish especially to acknowledge the help given by Professor Albert H. Sanford of La Crosse; Professor John D. Hicks of the University of California; the late Dr. Joseph Schafer, Superintendent of the State Historical Society of Wisconsin, and other members of the Society's staff, particularly Miss Livia Appel and Miss Alice E. Smith; Miss Sophia Sachs of Chicago; and my wife, Frances Clements Fries. De Paul University gave financial aid that made possible the completion of the book. I alone, however, am responsible for the points of view expressed and for any errors the book may contain.

The greater part of Chapter 9 was published in the December, 1948, issue of the *Mississippi Valley Historical Review* and is included here with the permission of the publisher.

The endsheet map is the work of Mr. Gunther W. Heller and Mr. Richard D. Houghton, Jr., of the Historical Society's staff. Mr. Houghton has also redrawn, from official records, the log marks that decorate the chapter openings. The rafting scenes facing pages 64, 65, 128, 129, and 225 are reproduced by courtesy of the H. H. Bennett Studio of Wisconsin Dells, Wisconsin.

ROBERT F. FRIES

Contents

List of Illustrations

WHi(X3)44754

Woodcut of a logging crew, probably produced under the Wisconsin Federal Art Project about 1935. Artist's name not recorded.

1. Introduction: The Setting

THE development of the United States illustrates impressively the elemental relationship between man and his environment which Ellen C. Semple phrased so neatly when she said, "What is today a fact of geography becomes tomorrow a factor of history." In this country the pervasive influence of rich physical resources and a generally beneficent climate has reached truly imposing proportions. From the very beginning it was the lure of natural wealth that attracted the explorer, the trader, and the colonizer, and from the earliest settlement to the present day the existence of untapped resources awaiting the magic touch of the enterprising developer has continued to condition and guide the destiny of America.

Until the United States in the recent past attained world eminence as the land of the machine, wood remained of all the gifts of nature the most characteristically American. One may almost say that the nation has been carved from wood. True, virgin soil and many varieties of minerals were the indispensable ingredients of this great agricultural and industrial civilization, but it was the vast primeval forests which supplied the framework for their efficient utilization. Beginning with the first white settlement on the Atlantic coast, the pioneers of each successive frontier, struggling to transform a tangled wilderness into an ordered civilization, made use of wood for houses and barns and fences, for fuel, for furniture and tools, for public buildings. When their struggle was rewarded with a measure of success and the confused mazes of nature gave way to bustling cities and broad, productive farms, their descendants still demanded wood, not only for homes but for factories and railroads, steamboats and carriages, and all the other appurtenances of the civilized state.

To the pioneer in the heavily wooded regions of the Atlantic seaboard and the trans-Appalachian country, the forests not only supplied a great necessity but they also presented, paradoxically, an obstruction to his progress. He was forced to do battle with his friend the forest, which furnished him the wood he needed, that he might have room for productive activity. And only slowly did the forest retreat before his axe. After two centuries of ceaseless effort he was still not its conqueror. The slowness of this advance into what seemed a limitless ocean of trees helped to implant in the American mind the illusion that the forests were inexhaustible. At the same time the obstructive aspect of the woods engendered a feeling of antagonism. These two concepts, both inimical to conservation, governed public policy and popular sentiment west of the Appalachians during most of the nineteenth century.

So long as the path of settlement hugged the forest area no extensive trade in lumber developed. The rural settler built his house of logs and his fences of rails; the village dweller received all the lumber he needed from the primitive sawmills that sprang up in every partially urbanized community. Only as the villages grew into towns and the towns into cities, as the local supply of wood vanished, did he find it necessary to buy lumber in more remote places. By 1830 the extensive forests of Maine, New Hampshire, New York, and Pennsylvania were being exploited on a large scale to supply the demands of urban centers on the seaboard. An important lumber industry had developed.

Meanwhile the agricultural frontier, like a huge tidal wave, swept over the Allegheny Mountains and pushed its way down the Ohio Valley. When the great tide of immigrants reached the treeless prairies of Illinois and broke across the plains west of the Mississippi, they created a demand for lumber such as America had not yet known. Great areas of these plains were extremely fertile. They offered the pioneer a rare opportunity to create a rich farm out of the wilderness without the traditional backbreaking toil of clearing his fields. But by the same token they did not furnish him the wood he needed for his buildings.

Fortunately it was no difficult matter to get the necessary wood. One of the most magnificent forests in America, connected with the unwooded area by easily accessible waterways, stretched across northern Michigan, Wisconsin, and Minnesota. Here was a situation truly made for men of vision and enterprise: on the one hand a seemingly inexhaustible supply of building timber, and on the other a seemingly endless expanse of treeless plain inevitably destined to be settled by the migrants who were restlessly surging westward.

Swallowed up by the voracious consumption of a growing population, the limited stands of timber in the middle Mississippi Valley had virtually disappeared before 1840, and a great cry for building materials went up from the disillusioned settlers. Their cry did not go unheeded. The few lumbermen who, more adventurous than the rest, had begun to operate in the Great Lakes forest area soon after 1800 increased steadily as news of the opportunity permeated the older logging regions of the East. As in the Eastern states, the first sawmills were small affairs, capable only of meeting the local demand. But they were the nuclei from which a new industry was to radiate as soon as extinction of the Indian title to the land had provided a firmer legal basis. The Great Lakes area was ready to supply not only the West but also the Eastern states with an abundance of lumber.

The woodlands which were to be the basis of the prosperity of the Great Lakes states for a half century and more were an extension of a vast forest beginning in eastern Canada and New England. Stretching far westward, and southward roughly to Lake Erie, this great stand of timber covered northern Michigan, Wisconsin, and Minnesota and finally tapered off in a northwesterly direction into western Canada. Originally, it is estimated, this forest contained some hundred billion board feet of commercial lumber, a wealth far surpassing all the fabled treasure troves of the world. In Wisconsin the true forest lay almost entirely north of a line drawn from Manitowoc to Portage and thence to the falls of the St. Croix River. It was not a homogeneous region. Its southern edge was a hardwood tract, from thirty-five to fifty miles wide, only five per cent of which was pine timber. North of it was a belt of mixed hardwoods and conifers, about equally divided in quantity. Within this zone were many open spaces, grasslands or marshes, comprising about five per cent of the area. Finally, along the headwaters of the Wolf, Menominee, Wisconsin, Chippewa, Black, and St. Croix rivers stood a growth of conifers.[1]

In the early years of the industry only the white pine interested the lumbermen of Wisconsin. Light and easily workable, it was an ideal wood for all building purposes and was easily transportable by water. It was because of this preoccupation with pine that the north woods came to be called "the pinery," though in few sections of it was pine actually predominant. Geologists designated as pineland any tract of several hundred acres which contained an average of one or two large pine trees to the acre and which was near a stream of sufficient depth to float logs to a mill. Both the quality and the amount of pine varied greatly from one region to another, often from one acre to the next. Some areas furnished as much as a million and a half board feet per forty-acre tract, almost forty thousand feet per acre. Others yielded only a thousand to thirty-five hundred board feet to the acre.[2]

The glacier which once covered most of Wisconsin had done much to help the lumber industry. Not only had it pulverized

[1] George W. Hotchkiss, *History of the Lumber and Forest Industry of the Northwest* (Chicago, 1898), 22; Mary Dopp, "Geographical Influences in the Development of Wisconsin," in the *Bulletin of the American Geographical Society*, 45:736–737 (October, 1913); Ray H. Whitbeck, "The Industries of Wisconsin and Their Geographic Basis," in *Annals of the Association of American Geographers*, 2(1912):59–60.

[2] *Geology of Wisconsin: Survey of 1873–1879* (official report of the state geologist, Madison, 1880), 3:326–327; Dopp, *op. cit.*, 738.

the soil of the northern part of the state into the sand upon which the white pine thrives, but it had also carved out numerous lakes ideal for storage and the streams upon which lumbermen depended for transportation before railroads were built into the forest.[3] Indeed, aside from the white pine itself, the rivers of Wisconsin were the most important factor in its lumber industry. Granted the need for improving them, they could hardly have been more useful to the industry had they been created expressly for lumbering. In the spring, reinforced by the melting of heavy snows, they gave great transporting energy. At frequent intervals they supplied ideal water-power sites for sawmills. They tended to flow outward radially, and at the boundaries of the state they joined the Great Lakes and the Mississippi River, which connected them directly with the great lumber marts of the East and West. So advantageous were these transportation facilities that the Great Lakes region supplied the Eastern market with lumber for many years before lumbering was begun in West Virginia. Before railroad transportation became general the rivers divided Wisconsin into six natural lumbering districts: the Green Bay and Wolf River districts in the northeastern part of the state, and the Wisconsin, Black, Chippewa, and St. Croix districts in the northwest.[4]

Thus by 1840 nature and the market economy had combined to set the stage for the lumberman in Wisconsin.

[3] Lawrence Martin, "Progressive Development of Resources in the Lake Superior Region," in the *Bulletin of the American Geographical Society,* 43:663 (September, 1911).

[4] Whitbeck, "Industries of Wisconsin," *op. cit.,* 59–60; Frederick Merk, *Economic History of Wisconsin during the Civil War Decade* (State Historical Society of Wisconsin *Studies,* vol. 1, Madison, 1916), 60–61; Hotchkiss, *Lumber and Forest Industry,* 380.

2. Vanguard of Conquest

DESPITE its obvious potentialities the lumber trade was only a minor activity of Wisconsin's first settlers. Not until the forties did it supersede the fur trade and lead mining, which had been the chief commercial enterprises. Throughout most of the thirties logging was done only on a small scale, in the face of a crying need for building materials in the prairie regions of southern Wisconsin, Iowa, and Illinois, the normal market area for pioneer lumbermen. In the lead-mining district around Mineral Point and Galena the miners who elected to remain throughout the winter had to dig into the hills for shelter. When the first territorial capitol was built at Belmont in 1836, the lumber needed for its construction had to be transported from a tributary of the Allegheny River in Pennsylvania down the Ohio and up the Mississippi to Galena, and from there by ox team. In that same year a Methodist missionary tore down his two-story house in Meadville, Pennsylvania, and shipped it to Prairie du Chien when he was transferred there. Settlers in northern Illinois and in Iowa paid as much as a hundred dollars for a thousand board feet of Allegheny River lumber or whitewood which had been laboriously carried by wagon train from the Wabash country.[1]

[1] John G. Gregory, ed., *West Central Wisconsin: A History* (4 vols., Indianapolis, 1933), 1:155, 159; Ella C. Brunson, "Alfred Brunson, Pioneer of Wisconsin Method-

This dearth of lumber may be attributed partly to the lack of capital and specialized labor, both of which were essential to the full exploitation of the pine timber of northern Wisconsin. And these two vital elements would not migrate in sufficient strength so long as the legal title to the forest areas of Wisconsin remained with the Indians, for federal law forbade white men to log on Indian lands without special permission. Even after the several tribes had ceded their lands by formal treaty, prospective loggers were expected to wait until the land was surveyed and sold.

Nevertheless some lumbering was done on a small scale before the government negotiated cession treaties with the Indians. There was little room in the frontier psychology for scruples about the free use of the public domain, and it was inevitable that some adventurous spirits should hazard the risks in order to take advantage of the nascent and growing demand for construction materials. The government itself, in building its military posts, took what timber it needed, and it recognized the futility of attempting to prevent others from doing likewise.

After 1830 it became a common practice of the secretary of war to issue permits to log on Wisconsin Indian lands. He had no statutory power to issue such licenses, and their recipients therefore assumed considerable risk of dispossession, whatever may have been their rights in equity. At best the permit was a negative concession, merely pledging that the War Department would not interfere with the stipulated logging operations so long as stated conditions were observed.[2]

With or without such permits, pioneer loggers conducted sporadic operations in Wisconsin from about 1810 until it became possible, after 1837, to place lumbering on an accepted business basis. Among them were Jacob Franks in the Green Bay district; Colonel John Shaw on the Black River; Constant

ism," in the *Wisconsin Magazine of History,* 2:138 (December, 1918); William R. Smith, *Observations on the Wisconsin Territory, Chiefly on That Part Called the "Wisconsin Land District"* (Philadelphia, 1838), as cited in Gregory, *West Central Wisconsin,* 1:158–159.

[2] "Some Wisconsin Indian Conveyances, 1793–1836," in *Wisconsin Historical Collections,* 15:11–15 (Madison, 1900).

Andrews, Hardin Perkins, James Lockwood, and James Rolette in the Chippewa River region; and Daniel Whitney on the Wisconsin River. These men can scarcely be credited with founding an industry, for most of their work produced no tangible results of a permanent nature. But they were the trail blazers who demonstrated to those who came later the exploitive possibilities of the Wisconsin timber areas.[3]

When it became apparent after 1830 that the government's failure to remove the red man from the pine forest of the Great Lakes region was impeding the settlement of the prairies to the south, the demand for the extinction of Indian titles rose to a clamor. The Indians still held title even to much agricultural land in the prairies, and the pressure of immigration was beginning to push considerable numbers of settlers into the Indian country. In 1832 the exasperated Sauk and Foxes precipitated the dreary campaign known as the Black Hawk War. Stung into action, the government now began to negotiate a series of treaties which in less than twenty years were to deprive the aborigines of all their lands in Wisconsin except a few square miles of reservation area.[4]

These first Indian treaties, coinciding as they did with the warm glow of Jacksonian prosperity in which the nation was basking, produced an immediate rush of immigrants to Wisconsin. By December, 1836, settlers and speculators had bought nearly nine hundred thousand acres of land. Eastern speculators, viewing the profits that lumbering was yielding in New York,

[3] For details of the careers of these and other pioneer lumbermen, see George Gale, *Upper Mississippi, or, Historical Sketches of the Mound Builders, the Indian Tribes, and the Progress of Civilization in the North-West; from A. D. 1600 to the Present Time* (New York, 1867), 354; Moses M. Strong, comp., *History of the Territory of Wisconsin from 1836 to 1848* (Madison, 1885), 80; Willard Keyes, "A Journal of Life in Wisconsin One Hundred Years Ago," in the *Wisconsin Magazine of History*, 3:453–457 (June, 1920); Hotchkiss, *Lumber and Forest Industry*, 474–476; Ellis B. Usher, *Wisconsin, Its Story and Biography, 1848–1913* (8 vols., Chicago, 1914), 1:188, 201; James H. Lockwood, "Early Times and Events in Wisconsin," in *Wisconsin Historical Collections*, 2:132–139 (Madison, 1856); Ebenezer Childs, "Recollections of Wisconsin Since 1820," *ibid.*, 4:175 (Madison, 1859); Albert G. Ellis, "Upper Wisconsin Country," *ibid.*, 3:437–438 (Madison, 1857).

[4] Louise P. Kellogg, "The Story of Wisconsin, 1634–1848," in the *Wisconsin Magazine of History*, 3:190 (December, 1919); Joseph Schafer, *A History of*

considered the Western timberlands a good investment, and many bought Wisconsin real estate, among them Daniel Webster, Edward Everett, Caleb Cushing, and Ralph Waldo Emerson. So feverish became the excitement over lands in the newly opened area that corner lots in the tiny village of Sheboygan were selling by map in New York for as much as fifteen thousand dollars. The panic of 1837 of course brought the wave of speculation to an abrupt end, but the lumber industry continued to grow, for it had not yet met the demand for lumber in the Mississippi Valley.[5]

The United States district attorney for Wisconsin Territory reported to Washington that as soon as the treaties of 1837 had been negotiated, "men of enterprise but generally without money" had begun seeking favorable sites for the establishment of logging and milling operations. By giving the "most flattering accounts of the advantages the country held out," the explorers managed to associate themselves with men of capital. Some ten or fifteen companies were thus established almost at once on the territory recently ceded by the Chippewa and Winnebago. These companies were small, most of them employing only fifteen to twenty-five men, but the district attorney was informed in 1838 that many new establishments would be started during the next season and the operations of the existing ones greatly enlarged. All this took place before the land had been surveyed, let alone put on sale.[6]

The growth of these early logging companies was contingent upon the migration to Wisconsin of men skilled in the tech-

Agriculture in Wisconsin (Wisconsin Domesday Book: General Studies, vol. 1, Madison, 1922), 134; Strong, *Territory of Wisconsin,* 216–218; Daniel S. Durrie, "The Public Domain," in *History of La Crosse County, Wisconsin* (Western Historical Co., Chicago, 1881), 218; Usher, *Wisconsin,* 1:47, 174; Ellis, "Upper Wisconsin Country," in *Wisconsin Historical Collections,* 3:438; William F. Raney, *Wisconsin: A Story of Progress* (New York, 1940), 76–79.

[5] Strong, *Territory of Wisconsin,* 216–218; William R. Smith, *The History of Wisconsin; in Three Parts, Historical, Documentary, and Descriptive* (Madison, 1854), 1:287; Schafer, *Agriculture in Wisconsin,* 30; Hotchkiss, *Lumber and Forest Industry,* 410.

[6] Moses M. Strong to H. O. Gilpin, November 20, 1838, in the Strong Lumber Papers. All private papers hereinafter cited are in the library of the State Historical Society of Wisconsin.

niques of logging, manufacturing, and transportation of the product. Most of the earlier millowners knew very little of the technical side of lumbering, and they soon faced problems they could not solve without expert help. The West could furnish many men skilled in woodcraft, but few who understood the technicalities of running an industry in the wilderness. Consequently the mill men sought to attract trained managers and laborers from the older lumbering communities of new Brunswick, Maine, New Hampshire, New York, and Pennsylvania.[7]

The Black Hawk War had helped to advertise the advantages of Wisconsin to Easterners in general, but the appeal for able lumbermen had to be more concrete. Paid advertisements were therefore published in New England newspapers and large amounts of promotional material circulated, all of which described in glowing terms the present prosperity of the territory, the potentialities of lumbering, and the vast amount of government land available at a nominal price. Newspapers in the pineries never tired of publishing articles enlarging upon the attractions of Wisconsin. Copied as they were by "exchange" papers, these pieces informed Eastern readers of the high wages being paid skilled workmen in the Wisconsin mills and woods, of the cheap lumber available for building purposes, and of the high prices paid for farm produce by the lumbermen. Much was made of the success of those who had already migrated to the territory. After 1850 a business depression in the Eastern pineries and the gradual exhaustion of the timber in the older lumbering areas also induced many skilled workmen to move west.[8]

[7] Isaac Stephenson, *Recollections of a Long Life, 1829–1915* (Chicago, 1915), 104. In 1848 a traveler reported that the mills around Big Bull Falls on the Wisconsin River were inactive for much of the year because there were not enough men in the woods to furnish an adequate log supply. "Report of J. G. Norwood, Assistant Geologist," in David Dale Owen, *Report of a Geological Reconnoissance of the Chippewa Land District of Wisconsin* (30 Congress, 1 session, Senate Executive Document no. 57, serial 509, Washington, 1848), 101–102.

[8] The following insertion in a Maine paper, the *Bangor Daily Whig and Courier* of October 9, 1838, is illustrative of the advertisements for skilled help: "Wanted to go West, one first rate head Sawyer, two of the second class, one who understands circular Saws, and one Teamster. The above to start immediately. Also in 3 or 4 weeks, a gang of 10 or 12 Wood Choppers. . . . The best of

The favorable publicity the Wisconsin lumbering districts received in the East produced immediate and lasting results. Virtually all the original lumber barons of Wisconsin—such men as Philetus Sawyer of Vermont, Isaac Stephenson of New Brunswick and Maine, the Doughty brothers of New Brunswick, and Orrin H. Ingram of Massachusetts and New York—had learned the business in the Eastern pine areas. Of the fourteen men who were recognized in 1898 as the leading lumbermen of the Chippewa district, four were natives of New York, three of Maine, two of Massachusetts, and three, respectively, of Pennsylvania, Canada, and New Jersey. Of the eight most prominent lumbermen of the Green Bay district two were born in New Brunswick, three in New York, two in Pennsylvania, and one in the Middle West. Small wonder that a congressman from Maine should have spoken plaintively of "the stalwart sons of Maine marching away by scores and hundreds to the piny woods of the Northwest." After 1840, moreover, the labor force from the East was supplemented by immigrants from the Scandinavian peninsula, Germany, and Ireland. In the half decade 1850–55 the population of the four principal logging counties more than doubled.[9]

Although men "generally without money" had been the first to enter the pineries, not many such entrepreneurs became successful millowners without the pecuniary assistance of others.

references will be required." Cited in Richard G. Wood, *A History of Lumbering in Maine, 1820–1861* (*University of Maine Studies,* second series, no. 33, Orono, 1935), 234. The promotional pieces are typified by such publications as the following: James S. Ritchie, *Wisconsin and Its Resources* (Philadelphia, 1857); Wisconsin Commissioner of Emigration, *Wisconsin* (n.p., 1853); and Martin Mitchel and Joseph H. Osborn, *Geographical and Statistical History of the County of Winnebago* (Oshkosh, 1856). For newspaper articles carrying the same message see the *Chippewa Union and Times* (Chippewa Falls), June 19, 1869, and the *Chippewa Herald,* April 26, 1873, and November 10, 1882. On the effect of depression and the decline of logging in the East see, for example, Stephenson's *Recollections,* page 81, and the *Chippewa Herald* of April 26, 1873.

[9] Hotchkiss, *Lumber and Forest Industry,* 300 ff. and 485 ff., from which the data on lumbermen of the Chippewa and Green Bay districts have been compiled; *Congressional Globe,* 32 Congress, 1 session (1851–52), Appendix, 389; *Annual Report of the Secretary of State of Wisconsin,* 1855, pp. 78–79. Even the businessmen who entered the Wisconsin pineries without previous experience in lumbering were for the most part immigrants from New England, New York, Pennsylvania, or New Brunswick. The outstanding exception was Frederick

Before lumbering could become a great industry in Wisconsin, it must have money as well as men from the East. Again the older lumbering communities were the best sources of supply.

Newspapers in Maine printed advertisements describing what a "rare chance" to make a fortune was being offered by the Wisconsin pineries to men having money to invest. Cincinnati, Buffalo, and Montreal papers carried notices of millsites for sale at La Crosse. Armed with letters of introduction from former New Englanders, Moses M. Strong, a man of many affairs in early Wisconsin, visited Maine in 1851 "for the purpose of enlisting the attention of Capitalists in what, he conceives may be made a very lucrative business on the Wisconsin River, namely the manufacture of Pine Lumber." Upon his arrival there he advertised in the *Bangor Daily Whig and Courier* for "some experienced, practical lumber manufacturer, of enterprise, industry and integrity" to join him as a partner in the lumber business.[10]

The legislature of Wisconsin also did its part to help attract capital when it set the legal rate of interest at twelve per cent and repealed the laws that called for forfeiture of the principal if a higher than the legal rate of interest was agreed upon by contract. In 1855 a writer boasted that Wisconsin laws gave "a freedom to the value of money unknown in any other State in the Union."[11]

Yet it was hard to hide the fact that the pioneer lumber capi-

Weyerhaeuser, a German immigrant. See Ellis B. Usher, "Puritan Influence in Wisconsin," in *Proceedings of the State Historical Society of Wisconsin*, 1898, p. 125; Joseph Schafer, "Yankee and Teuton in Wisconsin," in the *Wisconsin Magazine of History*, 6:125–145 (December, 1922); Wood, *Lumbering in Maine*, 228–229, 232; *Wisconsin Lumberman* (Milwaukee), 2:154 (May, 1874); *Eau Claire Free Press*, August 13, 1874; Usher, *Wisconsin*, 1:179; John E. Nelligan, *Life of a Lumberman* (Madison, 1929), 20–36; Stephenson, *Recollections*, 79. The most recent biography of a Wisconsin lumberman is Richard N. Current's admirable study of Philetus Sawyer, *Pine Logs and Politics*, recently published by the State Historical Society of Wisconsin.

[10] Stephenson, *Recollections*, 65; Wood, *Lumbering in Maine*, 226; *La Crosse Independent Republican*, January 16, 1856; letter of Charles Durkee to Joseph Grinnell, May 25, 1851, and clipping dated June 10, 1851, in the Strong Lumber Papers.

[11] Tom O. Edwards, "The Progress, Conditions, and Prospects of Wisconsin," in *Wisconsin Historical Collections*, 2:497 (Madison, 1856).

WHi(X3)1499

Moses M. Strong, a businessman prominent in the history of lumbering in Wisconsin (see Index under "Strong"). Steel engraving.

WHi(X3)44752

Vicinity of Cable, Bayfield County, Wisconsin, 1914. A load of logs coming down a grade at Ole Emerson's lumber property no. 4. Note the hay piled up along the grade and scattered in the grooves for the sled runners; this was needed to slow the motion of a heavy load on a downgrade since the horses effectively could not resist the push of the load behind them.

talist faced tremendous difficulties. One who today contemplates what a wealth of timber once covered the cutover wastelands of northern Wisconsin can easily exaggerate an earlier generation's opportunities to make a fortune. True, great opportunities did exist, and nineteenth-century businessmen were well aware of them. But they were also alive to the immensity of the task of piercing the wilderness, the seemingly endless obstacles to the establishment of a new industry far from a base of supplies, and the meagerness of the initial rewards. Of those who made the venture, many failed; in point of fact, during the early days of lumbering they exceeded the number of those who made a success of it. In any case one must not let hindsight obscure the fact that the very immensity of the forests led most people to take them for granted, much as they did the sunshine and the air about them.

Among the greatest problems of the early lumberman was the transportation of supplies into the pineries. Since few agricultural settlers had yet appeared to grow vegetables and grains, foodstuffs for the logging and milling crews had to be carried in from the outside by keelboats and tote teams over extremely difficult routes. Clothing, dry goods, ironwork, and mill machinery found their way into the pine woods in the same way. Transportation costs were therefore excessive, and foodstuffs exorbitantly high-priced. Between 1853 and 1857 pork cost from twenty-five to thirty-five dollars a barrel, and spring wheat flour from six to nine dollars a barrel. According to Isaac Stephenson, writing in 1915, these prices were higher than at any time thereafter.[12]

New concerns faced so uncertain a future, especially during hard times, that storekeepers refused to sell them supplies on credit. To overcome this handicap the more enterprising of the lumbermen entered the freighting business themselves and started stores of their own. After 1865, for example, the firm of

[12] B. W. Wilson & Co. to Dole and Ingram, December 22, 1857, and Dole, Ingram, and Kennedy to Barkley, Hale & Co., December 22, 1857, in the Orrin H. Ingram Papers; *Eau Claire Weekly Free Press,* May 19, 1870; Stephenson, *Recollections,* 118.

Ingram, Kennedy and Company, an Eau Claire lumbering concern, conducted the largest part of the freighting business from Read's Landing to Eau Claire, using a steamer built to their order.[13]

Even when the initial difficulties had been surmounted and a firm seemed to be on the way to making a fortune, the uncertainties inherent in the business could bring new hardships or even ruin. Early mills and power dams were often rickety affairs, imperiled by every rise in water level. Not infrequently a sudden flood swept away everything—mill, dam, and the entire supply of logs. Fire was a constant menace. Low water in the spring meant that few logs would reach the sawmill. Logging operations could be severely hampered by a subnormal snowfall in winter. In the days of river rafting every rapids, every dam, every pier and boom below a mill was a potential danger to its lumber on the way to market. Finally, when all mechanical difficulties had been conquered, all dangers surmounted, a financial panic and business depression, such as those of 1857 and 1873, might glut the market and depress the price of lumber to ruinous levels.[14]

All these hazards impeded lumber manufacturing in Wisconsin, but they could not prevent its steady growth. As early as 1843 the decline of the fur trade and of lead mining was forcing men out of these occupations into the new logging camps. Two years later, when Isaac Stephenson reached Wisconsin, lumbering had completely displaced mining and was becoming, with agriculture, the backbone of a new economy.[15]

Because of its greater accessibility to the older settlements and its wealth of pine resources, the Wisconsin River forest area was the first to be exploited on a large scale. In the years after

[13] John Strong to Moses M. Strong, May 24, 1849, in the Strong Lumber Papers; Henry Nazro & Co. to Dole, Ingram, and Kennedy, January 15, 1858, and Furton [?] and Sercomb to Dole, Ingram, and Kennedy, January 18, 1858, in the Ingram Papers; Hotchkiss, *Lumber and Forest Industry,* 494.

[14] John Delaney to Moses M. Strong, March 8, 1848, in the Strong Lumber Papers; Milwaukee *Daily Wisconsin,* December 24, 1857; Stephenson, *Recollections,* 91; A. M. Dole to O. H. Ingram, June 10, 1857, in the Ingram Papers.

[15] *Communication of the Hon. Alfred Brunson, Relating to His Travels in the North-Western Wilderness of Wisconsin* (Wisconsin Territorial House Document C, 1843, 12-page pamphlet), 11; Stephenson, *Recollections,* 66.

the panic of 1837, the very period when business in general was depressed throughout the country, the upper Wisconsin region enjoyed a tremendous boom. By 1839 early comers had occupied every available millsite as far north as Big Bull Falls, and by the end of the next year—so swiftly did operations expand northward—nearly every suitable site on the upper river had been taken.[16]

The lumber output of the district increased steadily for more than a quarter of a century. In 1840 markets down the river absorbed about six and a quarter million board feet of Wisconsin River lumber, the next year over seven and three-quarter million feet, and in 1842 eight and a half million feet. Activity was only slightly retarded by the Mexican War and the California gold rush. By 1847 twenty-four mills were producing nineteen and a half million board feet, worth about six dollars a thousand at the point of production, and three million shingles. Two years later a traveler found forty-seven mills in full operation, employing almost two thousand men to raft logs and lumber. Estimates of production in 1854 range from ninety-five million to one hundred and twenty-five million board feet. By 1857 the capital invested in lumbering on the Wisconsin had mounted to nearly six million dollars, and the value of its product probably exceeded four million dollars. This phenomenal growth continued during the Civil War and the years thereafter, and by 1872 the annual output of the Wisconsin River pinery had attained the enormous volume of two hundred million board feet.[17]

In the meantime other districts had been rapidly outstripping the Wisconsin River region. The length of the Wisconsin and

[16] *A Standard History of Portage County, Wisconsin* (Lewis Publishing Co., Chicago, 1919), 50; Walter A. Blair, *A Raft Pilot's Log: A History of the Great Rafting Industry on the Upper Mississippi, 1840–1915* (Cleveland, 1930), 35; Hotchkiss, *Lumber and Forest Industry*, 438.

[17] Owen, *Geological Reconnoissance of the Chippewa Land District*, 70; John E. Chapin, "Sketch of Cutting Marsh," in *Wisconsin Historical Collections*, 15:36 (Madison, 1900); Silas Chapman, *A Hand Book of Wisconsin: A Guide to Travellers and Immigrants* (Milwaukee, 1855), 5; John Gregory, *Industrial Resources of Wisconsin* (Chicago, 1853), 83; Mitchel and Osborn, *Geographical and Statistical History of the County of Winnebago*, 10; Hotchkiss, *Lumber and Forest Industry*, 438; Ellis, "Upper Wisconsin Country," in *Wisconsin Historical*

its numerous rapids made rafting on it an expensive operation, and the transportation of supplies to its pineries a difficult task. New lumber capital sought more profitable fields of investment.[18]

One of these was the Green Bay and lake-shore district, the swift growth of which had begun during the late forties. When Isaac Stephenson arrived in Wisconsin in 1846 he had found only two mills "of considerable size" at Green Bay, another at the mouth of the Menominee River, and four or five others scattered throughout the area. At that time, he estimated later, the total investment in lumber milling in the Green Bay area amounted to less than fifty thousand dollars. By 1854 various contemporary observers were estimating the output of the whole region, including part of Michigan, all the way from fifty-seven million to a hundred and thirty-five million board feet. By 1865 shipments from the area totaled two hundred million board feet of lumber besides shingles and lesser products. In 1871 the Wisconsin side of the Menominee alone was said to have produced three hundred million board feet of lumber.[19]

Meanwhile the Wolf River district was establishing similar records. For a few years after 1836 settlers around Oshkosh had received their lumber supplies from the two small mills that had been built for the use of the Brotherton and Stockbridge Indians, but by the mid-forties several other small mills had been built in the area to supply the local demand—two at Fond du Lac in 1844, and one at Oshkosh in 1847. An intense rivalry now sprang up between the two towns, which was to continue for several years. In 1846 enterprisers at Fond du Lac made a bid for a larger market by building a steam-powered mill, which is

Collections, 3:443; *Documents Showing the Annual Amount of the Trade and Commerce on the Upper Mississippi River* (28 Congress, 1 session, Senate Document no. 242, serial 434, Washington, 1844). No reliable method of gathering lumber statistics was developed in Wisconsin during the nineteenth century. The estimates of experts must be relied upon, although at times they vary so widely as to be useless, and at best they may be taken only as approximations.

[18] Merk, *Economic History of Wisconsin during the Civil War Decade*, 65–66.

[19] Stephenson, *Recollections*, 76, 142–143; Mitchel and Osborn, *Geographical and Statistical History of the County of Winnebago*, 10; Chapman, *Hand Book of Wisconsin*, 5; Hotchkiss, *Lumber and Forest Industry*, 291, 411, 418, 422, 429; Merk, *Economic History of Wisconsin during the Civil War Decade*, 65.

believed by many to have been the first of its kind in northern Wisconsin. Eight years later the city sold about seven million feet of lumber, mostly to neighboring farmers in exchange for wheat.

For a time Fond du Lac had the advantage in its race with Oshkosh. Its proximity to the port of Sheboygan brought it into closer touch with the Chicago and Milwaukee markets, and its rail connections with southern Wisconsin and northern Illinois were better than those of its rival. Oshkosh, however, being close to the mouth of the Wolf River, was nearer the source of supply and was able to use Lake Poygan for storage of logs, whereas Fond du Lac had to depend on logs rafted across Lake Winnebago. Moreover, Oshkosh was in a strategic position to ship its product by way of the projected Fox-Wisconsin Canal to the markets of the lower Mississippi, and it could supply southern Wisconsin and northern Illinois by way of the Rock River.[20]

In the end the advantages of Oshkosh proved to be superior. By 1852 it had three or four mills. Eight years later it was supporting twenty-five lumber dealers and manufacturers and two sash, door, and blind factories. By 1866 about fifty mills were in full operation, which in that year produced eighty-five million feet of lumber, eight million shingles, and about fourteen million lath. Their products were being shipped, in boats they themselves owned, to retail yards along the Rock River, and by other means to the wholesale centers of Chicago. This expansion was representative of the district as a whole, which in the years from 1854 to 1871 increased its production from some thirty or forty million board feet to a hundred and eighty million feet. There was "no class of people," boasted the writer of a piece of promotional literature, "in a more thriving condition than the lumbermen."[21]

On the Black River the cutting of square timbers and shingles

[20] Joseph Schafer, *The Winnebago-Horicon Basin (Wisconsin Domesday Book: General Studies,* vol. 4, Madison, 1937), 261, 267, 279, 285–286; Hotchkiss, *Lumber and Forest Industry,* 280, 385, 389.

[21] Oshkosh *Northwestern,* June 8, 1860, November 22, 1866, January 20, 1876; Mitchel and Osborn, *Geographical and Statistical History of the County of Winnebago,* 10, 117; Chapman, *Hand Book of Wisconsin,* 5; Merk, *Economic History of Wisconsin during the Civil War Decade,* 65–66.

by hand for sale down the Mississippi had become an important business by 1840. In 1839, the year after the Indian title to the region had been extinguished, Jacob Spaulding and the Wood brothers, Robert and Andrew, had led an expedition of twenty men from Prairie du Chien to the falls of the river, where they built the first successful mill in the area. Four years later some three million feet of lumber was carried down the river. By 1844 the number of mills had increased to eight and by 1847 to thirteen, in which year they produced over six and a quarter million board feet.

La Crosse, located strategically at the confluence of the Black and La Crosse rivers with the Mississippi, became the natural focal point of the lumber trade of the region. The first mill there was built in 1852. Only a year later the town had ten sawmills, three shingle mills, and a sash, door, and blind factory. Estimates of the Black River output in 1854 ranged from eighteen million to forty-eight million feet, but at the same time La Crosse complained that it lacked lumber for its own building activities. Presumably the scarcity was only temporary, however, for it was expected that in 1855 the fourteen steam mills within a four-mile radius of the boat landing at La Crosse would produce fifty million feet of lumber. In 1859 the editor of a La Crosse newspaper proclaimed ecstatically that the city was destined to become the largest lumber mart on the Mississippi River. Stimulated by the high prices of the Civil War period, the industry on the Black River continued to prosper, and by 1872 was producing annually three hundred million board feet.[22]

But it was the Chippewa River valley, which contained about a sixth of the pine timber west of the Appalachians, that

[22] *La Crosse Tribune and Leader-Press,* August 2, 1930, July 19, August 8, 30, September 6, 27, October 4, 1936, April 18, 1937, February 20, 1938; *Communication of the Hon. Alfred Brunson,* 4; Gregory, *Industrial Resources of Wisconsin,* 83; Owen, *Geological Reconnoissance of the Chippewa Land District,* 70; Usher, *Wisconsin,* 1:174, 184; Ritchie, *Wisconsin and Its Resources,* 118; Mitchel and Osborn, *Geographical and Statistical History of the County of Winnebago,* 10; Chapman, *Hand Book of Wisconsin,* 5; *La Crosse Independent Republican,* August 23, 1854, July 25, 1855, December 31, 1856, May 25, 1859; Merk, *Economic History of Wisconsin during the Civil War Decade,* 65-66.

rapidly became the most important area of the Wisconsin lumber industry. By 1872 a mill on the Red Cedar which had been built by James Lockwood in 1828 and acquired in 1846 by John H. Knapp and William Wilson had become what purported to be the largest establishment of its kind in the country. Other millsites along the river were soon occupied, and the lumber output mounted rapidly. In the years 1843–55 it increased from five and a half million feet to seventy-five million feet. During 1855 and 1856 an exciting boom was precipitated by speculators, lumbermen, and settlers, who indulged in a mad scramble at the Eau Claire land office. Then came the crash of 1857, in which many lumbermen were ruined. But the Civil War again restored prosperity, and in the decade 1860–71 the annual log cut increased from sixty million to four hundred and thirty-six million feet.[23]

The output of the St. Croix district, part of which must be credited to Minnesota, also mounted rapidly after 1840. In 1843 eight million feet of lumber was produced, in 1847 seven and three-quarter million feet, and in 1855 one hundred and sixty million feet. By 1864 six hundred thousand dollars was invested in logging operations alone, and almost fourteen hundred men were employed in cutting timber. In the spring of 1865 the river carried more than two hundred million feet of logs to the sawmills, and in 1872 mills on the Wisconsin side alone produced more than a hundred million feet. Many St. Croix logs were processed at mills on the Mississippi.[24]

The burgeoning lumber industry of the state is more dramatically reflected in the attitude of Wisconsin's leaders than

[23] *Trade and Commerce on the Upper Mississippi;* Gregory, *Industrial Resources of Wisconsin,* 83; *Eau Claire Free Press,* July 16, August 13, 27, 1874, November 1, 1883; Mitchel and Osborn, *Geographical and Statistical History of the County of Winnebago,* 10; Chapman, *Hand Book of Wisconsin,* 5; *La Crosse Independent Republican,* July 9, 1856; Merk, *Economic History of Wisconsin during the Civil War Decade,* 65–66.

[24] *Trade and Commerce on the Upper Mississippi;* Gregory, *Industrial Resources of Wisconsin,* 83; Chapman, *Hand Book of Wisconsin,* 5; Mitchel and Osborn, *Geographical and Statistical History of the County of Winnebago,* 10; *La Crosse Independent Republican,* July 9, 1856; *La Crosse Weekly Democrat,* February 9, 1864, May 29, 1865; Merk, *Economic History of Wisconsin during the Civil War Decade,* 66.

in the cold statistics of production. As early as 1850 the lusty young lumber capitalists, conscious of their growing importance, were demanding in typical Western fashion that the federal government grant them recognition in the form of more frequent mail deliveries to the pineries. They asked also for a more generous federal land policy, a cause which in 1852 moved Representative Benjamin C. Eastman of Wisconsin to deliver in Congress a ringing eulogy of his lumbermen constituents. They were, he said, "a class of persons who fear no danger," a group who had "with intense labor . . . penetrated the forests, and . . . opened a trade in lumber, which at the same time supplied the settlers upon the vast prairies of Illinois, Iowa, and Missouri, and the cities and towns upon the Mississippi with lumber, and opened a market for the corn, flour, beef, and cattle of the farmers which they took in exchange." A year later the Wisconsin commissioner of emigration announced exultantly that Eastern lumber was no longer competing with Wisconsin pine on the Mississippi. "The course of the lumber trade may now be considered as permanently changed," he wrote. "The pineries of Wisconsin now control, and will soon hold exclusive possession of the markets of the valleys of the Mississippi and its great western affluents."[25]

The rapid growth of a network of Middle Western railroads in the decade after the Civil War set the stage for the final steps in the industrialization of the lumber trade in the Great Lakes region. For it was inevitable that in the wake of "the cars" should follow speedy settlement of the plains west of the Mississippi, nationwide distribution of raw materials, and mass production. At first the lumbermen had only a dim realization of what their industry was to become, but the Civil War revealed its ultimate goal. The spurt the lumber business took during

[25] Moses M. Strong to J. Collamer, February 19, 1850, in the Strong Lumber Papers, indicating that lumbermen especially desired semi-weekly delivery on the route from Fort Winnebago to Plover Portage, which connected the pineries with their markets; *Congressional Globe,* 32 Congress, 1 session (1851–52), Appendix, 851; Wisconsin Commissioner of Emigration, *Wisconsin,* 7. The passage quoted was copied by *Hunt's Gazetteer* and was widely circulated in the East. Usher, *Wisconsin,* 1:156–157.

the last two years of the war and the rapid shift of population to the West in the years thereafter placed lumbering upon a large-scale basis that was to last as long as the timber upon which it was founded.[26]

[26] Ernest Bruncken, *North American Forests and Forestry: Their Relation to the National Life of the American People* (New York, 1900), 54–55; Matthew G. Norton, *The Mississippi River Logging Company: An Historical Sketch* (n.p., 1912), 9; Merk, *Economic History of Wisconsin during the Civil War Decade,* 59. The Civil War, says Jenks Cameron, "was a preparation of the national mind for what was to come by means of a graphic demonstration of grand scale activities." *Development of Governmental Forest Control in the United States* (Baltimore, 1928), 101.

3. Gathering the Raw Material

IF Germany had become "the teacher of the world" in the field of forestry, American lumbermen had become "the most expert exploiters of the natural forest resources," observed the chief forester of the United States in the nineties. "Methods of cutting, handling, sawing, marketing, and all the appliances and tools employed have been developed to the highest degree and all means have been adapted to the end which from the viewpoint of private interest appears desirable, namely, largest immediate profits." And of these various branches of the industry probably none had a greater effect upon net profits than the work done in the woods, the cutting of the trees and their preparation for transportation to the mill. It was here that superior skill paid the highest dividends.[1]

The organization and efficient operation of a lumber camp demanded a rare combination of managerial ability, technical skill, and woods lore. The camp boss was often called upon to perform, besides his duties as supervisor of the crew, the functions of physician, storekeeper, diplomat, banker, and law enforcement officer. Needless to say, men capable of conducting logging operations successfully under these conditions were not easy to find, though in prosperous times many were ready to try

[1] Bernhard E. Fernow, "American Lumber," in Chauncey M. Depew, ed., *1795–1895: One Hundred Years of American Commerce* (2 vols., New York,

their hand at it. Many firms went bankrupt simply because of the inefficiency of their logging camps, and ultimately most companies preferred to assign a large part of their logging operations to contractors.

In the early days all logging was done in winter, when ice and snow facilitated the transportation of the logs from the stump to the riverbank. The foreman's responsibilities began in the fall, usually sometime in September, when he examined the area that was to be logged and selected the site for the camp. Noting the salient features of the terrain, its slope, and the location of the stream which, come spring, was to carry the logs to the mill, he tried to find a well-drained spot near the center of the tract to be cut. With a skeleton crew he now laid out the main road and its branches so that every section of the tract would have convenient access to a road over which logs could be hauled to the stream. In the meantime his crew was also constructing the camp buildings.

Before 1860 logging camps were small and primitive affairs. One building provided sleeping and eating quarters for the entire crew. Its architecture was dictated more by expediency than by ideals of beauty or durability. Constructed, naturally, of logs, its low walls encompassed a space barely large enough for the purpose it was to serve. The roof was composed of "shakes," thin pieces of pine wood, which were held in place by poles plastered over with clay. Inside, in the middle of the earth floor, a low mound of sand called the caboose provided a place for a fire, the smoke from which escaped through a hole in the roof. In the more primitive camps the men slept on the floor, huddled together for warmth. Oftener they slept in bunks built of rough lumber around the walls. The "deacon's seat," a bench extending from the lower tier of the bunks, was the focal point of the

1895), 1:201. "In sawing which is essentially a matter of mechanical equipment and arrangement, conditions are everywhere equal and no mill enjoys an advantage over competing mills, but in logging there is a wider field for the expenditure of individual effort and exercise of skill and it is in this that the profits are made or losses of operation sustained." Stephenson, *Recollections,* 84–85. Logging methods varied slightly from place to place and from time to time, but in the main the description given in this chapter applies equally well to all parts of the state.

crew's social life. The earliest lumber camps had only a stable besides this main structure, and perhaps a smaller building for the manufacture of shaved shingles.[2]

As logging operations expanded and the advent of the logging railroad enabled lumbermen to build more permanent quarters, the camps gradually became more imposing. In general the plan of the sleeping quarters remained the same, but the building itself was enlarged, and a stove replaced the open fire. In time it became necessary to build more than a single cabin for the men. The number of stables increased also, and buildings were erected for other purposes. The most notable of these was the "wanigan," or company store, where the crew could procure clothing and tobacco, sometimes at exorbitant prices, and charge them against future wage payments. Usually the foreman and his clerk slept in the building housing the wanigan. By the 1880's the average camp included blacksmith and carpenter shops, a granary, hayshed, and, frequently, a kitchen and mess hall separate from the bunkhouse.[3]

Equally important were the logging roads of the lumber camp. A supply road, usually called the "tote road," provided communication with the outside world. In addition it was necessary to grade roads over which logging sleds could carry their loads to the riverbank. In the large camp these formed a complex network, the upkeep of which was a constant problem. At intervals, where the road came closest to the trees being cut, skidways were constructed to facilitate the loading of the logs upon the sleds. The skidway consisted simply of two long timbers placed parallel with each other at right angles to the road, the small ends of which were buried in the ground and the larger ends elevated on a head block to bring them level with the bed

[2] Bruno Vinette, Some Reminiscences of a Veteran Logger in Northern Wisconsin, pp. 3–4, manuscript dated 1923 in the possession of the State Historical Society of Wisconsin; *Chippewa Herald,* October 28, 1871; Oshkosh *Northwestern,* December 30, 1880.

[3] Ceylon C. Lincoln, Lumbering in Early Days in Wisconsin, p. 2, manuscript dated 1915 in the possession of the State Historical Society of Wisconsin; *Eau Claire Free Press,* February 5, 1885; W. W. Bartlett to Ruth Bartlett, January 18, 1921, one of a group of copies of letters written from camps of the Park Falls Lumber Company and the Hines Lumber Company, in the possession of the State Historical Society of Wisconsin.

WHi(X3)24628

The *Dancing Annie*, probably a lumber company "wannigan" or floating office and supply store, used by the Chippewa Lumber & Boom Co., near Eagle Rapids, Wisconsin, about 1900. Photo probably by Charland.

WHi(X32)6308

Cook and cookee (cook's assistant) in the kitchen shanty of a lumber camp. The cook was an extremely important man in the workings of a lumber camp and exercised absolute control over his domain, so long as he produced good food and plenty of it.

of the sled. The efficiency of the logging roads was increased in the 1870's when an inventive logger developed a method of icing them. In the fall before the ground became frozen two ruts were gouged in the road with instruments resembling a carpenter's plane which were attached to the hind runners of a sled. When freezing weather arrived, water was poured into them from a tank carried over the road by a sleigh. After the water had frozen, a single team could haul a load of incredible size over roads prepared in this way.[4]

Having laid out the roads and constructed the camp buildings, the foreman and his skeleton crew began building flooding dams in the driving streams wherever they were needed to insure a constant supply of water to carry the logs to the mill in the spring. In later years they were often built cooperatively by all the lumbermen using a particular stream or by the federal government.

By the time freezing weather arrived, the foreman was ready to complete his logging crew and arrange for his winter's supplies. Before 1860 the ordinary logging crew consisted of no more than twelve or fifteen men, but by the eighties it commonly numbered at least sixty, and sometimes a hundred or even more.

Each man had well-defined duties. Next to the foreman, the cook was perhaps the most important functionary in the camp, for substantial and palatable food was essential to men who performed the arduous labor required of a logging crew. A camp with a poor cook was likely to have "what was called two crews, one a coming and one a going." The actual felling of the trees was done by "choppers." Sawyers cut them into logs of proper length for the mill. Skidders supervised the dragging of the logs from the stump to the skidways. Swampers kept the roads in repair and cut away brush for the skidders. Teamsters drove and cared for the work animals. A clerk kept the wanigan records and tallied the "time" of the men. Most camp crews also included a scaler, who measured the amount of logs piled at the

[4] *La Crosse Tribune and Leader-Press,* March 13, 1938; Lincoln, Lumbering in Early Days (MS.), 3; *Eau Claire Free Press,* January 30, 1879.

landings; and the larger camps employed a blacksmith, a carpenter, a saw filer, and perhaps other artisans to keep the equipment in repair. Ordinarily the camp crew was divided into smaller contingents, each composed of men doing the same kind of work.[5]

Until 1880 the draft animal most commonly used in lumber camps was the ox. The early camps, composed of a dozen men or so, needed only three yoke of oxen, but the larger camps of course used more. Some loggers owned their work animals; others hired them of farmers, who often came long distances to find winter work in the woods. The oxen owned by the lumber company itself were usually left in the woods during the summer months to forage for themselves.[6] After iced roads came into general use the horse largely displaced the ox.

To get supplies to loggers in the woods was always a problem for the entrepreneur. The first lumbermen had to rely on keelboats to carry food and goods even to the site of their mills. This was in itself a tedious, expensive process, and there still remained the arduous task of moving the heavy supplies by wagon and sleigh over the tote roads into the woods. In 1876 fourteen teams were kept busy hauling supplies to the camps of a single Eau Claire company. It too was an expensive operation. In 1863, for instance, it cost a dollar and seventy-five cents to carry a hundred pounds of food or tools from St. Croix Falls to Spooner, a distance of about seventy-five miles. This particular item of expense was reduced to less than forty cents when railroad transportation became available, but of course that was not everywhere. Economy-minded lumbermen tried various expedients to reduce the cost of supplies. Some made the experiment of driving live beef to their camps. In 1860, when prices in general were low, it cost only ninety dollars to send eighty-one head of

[5] Lincoln, Lumbering in Early Days (MS.), 1–3; *Eau Claire Free Press,* February 5, 1885; Frank Hartman, "Life in a Lumber Camp," in *La Crosse County Historical Sketches,* Series 3: *The Lumber Industry* ([edited by Albert H. Sanford], La Crosse, 1937), 18–23.

[6] *Eau Claire Free Press,* February 15, 1883; Vinette, Reminiscences (MS.), 3; Lincoln, Lumbering in Early Days (MS.), 3; Otis W. Terpening to Charles E. Brown, March 1, 1931, in the Brown Papers; *La Crosse Tribune and Leader-Press,* March 13, 1938; Nelligan, *Life of a Lumberman,* 42, 117–118; George O. Jones, Norman S. McVean, and others, eds., *History of Lincoln, Oneida, and Vilas Counties, Wisconsin* (Minneapolis, 1924), 176.

cattle from Fort Madison, Iowa, to Menomonie, a distance of more than four hundred miles. Others established their own freighting business with some success. Nearly every lumberman whose operations were fairly extensive discovered that the profit earned by his own supply store could partially offset transportation costs. Properly managed, a store could increase the lumber company's net profits considerably.[7]

The activities of a lumber camp were governed by a strict routine. At about half past three in the morning the cook and his assistant, the "cookee," piled out of bed. While the cook began making breakfast, the cookee built fires in the kitchen, the bunkhouse, and the office. At four o'clock he wakened the teamsters, who immediately fed and harnessed the animals in the barns. The rest of the crew rolled out of their warm bunks at about half past four, and fifteen minutes later were responding to the cookee's tin dinner horn or his yell that breakfast was on the table. The meals were substantial in bulk, if somewhat lacking in variety.[8] Immediately after breakfast the teamsters, loaders, and landing men set off for work. The rest of the crew waited for the first rays of daylight before following them, but before the sun was fully visible the camp was in full action.

The primary job of the camp crew was of course the felling of the trees. At first only the choicest pines growing close to the stream were cut. Some of them were large enough to produce from four to five thousand board feet of lumber. As the best ones were taken, the loggers moved farther away from the banks of streams and began cutting smaller trunks. By 1874 a tree producing three hundred board feet was considered average, and before the industry had run its course in Wisconsin the diameter of a merchantable log had dropped from fourteen to five inches. As prices rose and smaller logs became marketable a piece of land could be cut over a second and a third time. One tract be-

[7] Vinette, Reminiscences (MS.), 1–2; *Northwestern Lumberman* (Chicago), November 4, 1876; Edward W. Durant, "Lumbering and Steamboating on the St. Croix River," in *Minnesota Historical Collections*, vol. 10, pt. 2 (St. Paul, 1905), 646; John H. Knapp Diary, 1860, manuscript in the possession of the State Historical Society of Wisconsin; *Chippewa Falls Democrat*, September 2, 1869; Oshkosh *Northwestern*, March 7, 1872.

[8] *Eau Claire Free Press*, February 5, 1885; Nelligan, *Life of a Lumberman*, 121.

longing to the Wisconsin Central Railroad paid the third man who cut over it as great a profit as the first. Moreover, land that had once been logged for white pine soon became valuable for its Norway pine, hemlock, oak, or other hardwoods.[9]

No tool of the lumberman, with the possible exception of the saw, better symbolizes the industry than the axe, "the emblem of pioneership, and the precursor of settlement and development."[10] The axes used in the logging camps varied with the task they had to perform. A thick, heavy bevel facilitated the felling of large trees by serving as a wedge with which to force out a chip with the same stroke that made the incision. To chop down a tree a variety of skills was needed, but the frontier abounded in expert axemen who knew how to fell a tree in any given direction.

That is not to say that they escaped their share of the dangers surrounding the lumberjack. It was the practice in felling a tree to cut a notch with axes on the side of the trunk toward which it was expected to fall. The sawyers would then begin to saw from the other side a little distance above the gash, inserting wedges at the proper time to cause the tree to fall. At a certain point in this process the weight of the tree might produce a longitudinal crack that would not only ruin the trunk but place the sawyers in deadly peril if it was not checked at once. Even an expert could not foresee in which direction the crack would run, and if it was not checked the entire upper part of the tree might descend perpendicularly upon the men below. If this danger passed and the tree began to fall properly, there was still a chance that it might strike the branches of other trees and be unexpectedly deflected from the direction chosen for its fall.[11]

When a tree had fallen, its limbs were trimmed off, and the sawyers began cutting it into suitable lengths for the mills, usually about sixteen feet but sometimes less, twelve or fourteen

[9] *Chippewa Union and Times,* March 2, 1867; *Eau Claire Free Press,* March 10, 1870, January 22, 1874, March 8, 1883; *Eau Claire Leader,* January 29, 1922; Memorandum for Cutting Logs, October 4, 1910, in the Holt Lumber Company Papers; Dopp, "Geographical Influences in the Development of Wisconsin," in the *Bulletin of the American Geographical Society,* 45: 743–744.

[10] *Eau Claire Free Press,* October 25, 1883.

[11] Charles E. Russell, *A-rafting on the Mississip'* (New York, 1928), 61–62.

feet. Meanwhile the swampers were clearing a trail through the brush to one of the tributary logging roads. Several methods were devised for "snaking" a log to the skidways, the most primitive being to attach a chain to it by which a yoke of oxen could drag it over the snow. A more common method was to raise the forward end of the log by a device known as a "go-devil." Many camps adopted an instrument resembling a pair of ice-tongs that could be clamped onto the end of the log.[12]

At the skidway the logs were piled with a block and tackle, preparatory to loading them on sleds. A logging sled consisted of two pairs of runners with a "bunk" on each set, which together constituted a bed for the logs. In the early days of the industry the bunks were usually six feet long, but in time their length was increased, to ten, twelve, and even twenty feet. In loading a sled the first log was placed in the middle at right angles to the bunks, and the next two were chained to the sides. The space between was then filled, and logs continued to be piled tier upon tier until the load taxed the capacity of the sled, the road, and the horses. The whole was held in place by binding chains. The loads were often of enormous size, and a loaded sled coming down the road was a veritable "moving building." A single team would pull from five to twelve, even seventeen, thousand board feet, and it was not unusual for camps to vie with one another over the size of their loads.[13]

One may imagine what skill was required to handle a team pulling such a load. Even with iced roads it was sometimes necessary to use pulleys and cables to overcome inertia and get a sled started. Once on its way, a single team could drag a huge load fairly easily on a level iced road, but there were other problems. At a steep upgrade an extra team, stationed there for the purpose, often had to supply more power. On an abrupt downgrade there was always danger of a runaway and perhaps serious catas-

[12] *Eau Claire Free Press,* October 29, 1885; Otis W. Terpening to Charles E. Brown, December 6, 1931, in the Brown Papers.

[13] T. T. Babbitt, Comparison of the Lumber Industry of Wisconsin with That of the Northwest, p. 1, manuscript dated 1921 in the possession of the State Historical Society of Wisconsin; Lincoln, Lumbering in Early Days (MS.), 3; Oshkosh *Northwestern,* March 7, 1872; *Chippewa Union and Times,* March 2, 1867; *Eau Claire Free Press,* January 4, 1877, April 30, 1885.

trophe. To forestall such accidents a workman called a "road monkey" sprinkled sand over the ice early in the morning, or perhaps covered it with hay. Where the grade was especially steep, the road monkey remained at his post all day, covering the ice anew after each sled had gone by.[14] When it arrived at the landing on the riverbank, the sled was unloaded and the logs "banked" on rollways until the spring thaw opened the stream and they could be floated to the mill.

This piling of the logs on the rollways was perhaps the most dangerous operation of all. They were heaped high on the riverbank close to the water's edge or even upon the ice itself so that in the spring they could be started downstream as easily as possible. As the logs were lifted into the air by a block and tackle or a crane, a workman called the "top-decker" stood on the pile to steer each one to its proper place. If the log on the end of the rope swung too far or too fast, it could easily mangle him or knock him off the pile to his death. Such accidents were "so common that when word went around the camps that another 'top decker' was dead no one asked how did he die but only where was he crushed."[15]

HEWING Timbers

Pioneer lumber camps usually converted part of their logs into hewn timbers or handmade shingles. The hewing of a timber was a task in which the expert axeman reveled. After the tree selected for hewing had been cut down, the choppers cleared limbs and bark from the portion that was to be hewn. When the lines of the prospective timber had been marked off, a workman mounted the log and with a broadaxe split away the excess wood. Expert wielders of the broadaxe from Canada and Maine, many of whom had worked in shipyards, could hew a timber almost as smooth as it could be planed. When two sides had been hewn, it was rolled over on one of its flat sides, and the other two sides were hewn. Sometimes it was finished with a huge plane, called a rosser, which was weighted with stones and pulled by horses. Expert hewers were likely to be particular about the type of axe

[14] Babbitt, Comparison of the Lumber Industry of Wisconsin with That of the Northwest (MS.), 2; Lincoln, Lumbering in Early Days (MS.), 5; Oshkosh *Northwestern,* March 7, 1872.

[15] Russell, *A-rafting on the Mississip',* 62.

they used, and logging companies usually had a variety of tools on hand from which the axeman could make his choice.[16]

Shingles were made by hand in a building constructed for the purpose. Taking a pine log of particularly straight grain, the shingle maker sawed it into blocks, which he then split into thin pieces with an instrument called a "frow." After being split the shingle was laid in a warm place until it had thawed, after which it was placed in a vise and tapered with a drawknife. An expert could shape three thousand shingles a day, but two thousand was considered a good day's work.[17]

The long workday came to an end only when it had grown too dark to see clearly. Last to arrive at the camp at nightfall were the teamsters, loaders, and landing men, who must wait until all loads had been properly banked. During a mild winter many camps began hauling logs at midnight to take advantage of the day's lowest temperatures. Illumination was supplied by oil-soaked rag torches set in tin reflectors, which were mounted on poles and placed at intervals along the roadways. When spring came unseasonably early, many loggers were forced to work day and night to get their quota of logs banked before all the snow and ice disappeared.[18]

Time produced many changes in the tools and techniques of logging. In the half century after 1845 everything connected with the lumber industry increased in size and efficiency. The simplest objects were improved in the interests of economy. The straight axe handle gave way to the curved handle. An improved crosscut saw eliminated the axe as a felling tool and enabled two men to produce in a day as many logs as four men had formerly cut. Steam dragsaws and tree-fellers were introduced before 1900, but they were not widely used. In 1889 many loggers experimented with a steam log-hauler which it was hoped would supplant horses and oxen in the woods, but it

[16] Vinette, Reminiscences (MS.), 1–2; Augustus Cole to Holt and Balcom, June 16, 1887, in the Holt Lumber Company Papers.

[17] Vinette, Reminiscences (MS.), 4–5.

[18] Nelligan, *Life of a Lumberman,* 129; *Eau Claire Free Press,* February 5, 1885; Bruncken, *Forests and Forestry,* 83–84; *Northwestern Lumberman,* March 3, 1877; La Crosse *Chronicle,* March 14, 1889.

proved unsuccessful. By 1900 horses were being used almost exclusively in the woods.[19]

The most revolutionary of all changes were those brought about by the railroad. The temporary narrow-gauge roads which were built from the logging tracts to the mill eliminated the need for banking on a stream and made it feasible to cut timber in the summer months. As early as 1866 summer logging was being carried on around Shawano, and by 1890 it was becoming fairly common in all the logging regions. Summer logging was further facilitated when lumbermen learned to drag their logs from stump to railway on greased runways. The railroad also promoted the logging of hardwoods, which in time came to be in almost as great demand as white pine, but which were likely to sink in the driving streams. After being cut into short lengths in the forest, the hardwoods were hauled to the nearest railroad on sleds and sent to the mill by rail.[20]

As the desirable pine near the streams was exhausted, longer hauls were required, which in turn necessitated the use of larger sleds. Some of the logging sleds in use during the nineties weighed as much as five thousand pounds and had runners as long as nine feet. Such vehicles could not run on the hastily prepared logging roads that had been common in the seventies, and intensive road-grading operations became necessary. After being graded by heavy breaking plows and wheel-and-drag scrapers, some of these main logging roads were good enough to serve as railroad rights-of-way.[21]

Many improvements in logging technique were adopted by

[19] Fernow, "American Lumber," in Depew, *American Commerce,* 1:201; William H. C. Folsom, *Fifty Years in the Northwest* (St. Paul, 1888), 706; Nelligan, *Life of a Lumberman,* 117; *Chippewa Herald,* January 28, 1898. The builders of the steam haulers contracted with the loggers to supply the machines and the men to run them for a fixed rental. When the experiment failed, the loggers who had depended on the machine to get their logs banked sued for damages. Glover and Chandler to Orrin H. Ingram, October 19, 1888, in the Ingram Papers; *Chippewa Herald,* January 25, March 15, 29, 1889.

[20] Oshkosh *Northwestern,* July 5, 1866; La Crosse *Chronicle,* April 13, 1882; *Chippewa Herald,* November 21, 1890, April 22, 1898; Nelligan, *Life of a Lumberman,* 41; Otis W. Terpening to Charles E. Brown, December 6, 1931, in the Brown Papers; Stephenson, *Recollections,* 86.

[21] C. L. Tolles, "Winter Logging in the North Woods," in the *Lumber Worker* (Nashville, Tennessee), 4:47–55 (February, 1928).

contractors who were forced by cutthroat competition to increase their efficiency or be ruined. As the scale of lumbering operations became ever larger, owners of timberland and mills let an increasing amount of their logging work to contractors, who assumed all the risks and furnished their own equipment and labor. In prosperous times such log jobbers, requiring comparatively little capital, multiplied almost overnight. When the hard times following the panic of 1873 drew to a close, one newspaper noted that "nearly every taylor and shoemaker, from Manitoba to St. Louis have made up their minds to quit their benches and cut logs this winter, and have been finding chances to log and men to supply them, or guarantee their bills.... The old loggers, have nothing else to do, are bound to go in and squeeze the new men, trusting to Providence to bring them out with the bills for the pork and beans paid, and leaving the laborers to get their pay under the thieving lien laws."[22] Such conditions inevitably produced disastrous competition among the loggers, whereas the other parties to the logging contracts found themselves in a favorable position.

There was no uniformity in the logging contracts. Indeed, they varied as much as the circumstances under which they were drawn. A mill company might own the land from which it received its logs, it might own only the stumpage, or it might own neither. Sometimes the jobber himself owned the land or the stumpage upon which he worked. Sometimes a third party owned it. Sometimes a logging contract was let by a person who owned neither land nor mill but hoped to sell his logs to a millowner after they were cut. A landowner might contract to have the pine cut from his holdings and later sell the logs to a mill, or he might have the mill saw them on a share basis. Absentee owners frequently contracted with resident lumber companies or independent loggers to cut their logs as an alternative to outright sale of the land. If the agreement was made with a mill company, the contract usually provided for the sawing and marketing of the product.

[22] *Ibid.*, 55; Bruncken, *Forests and Forestry*, 86–87; Oshkosh *Northwestern*, November 7, 1878.

Most commonly the logging contract was let by a millowner, and it stipulated that the jobber cut and deliver a specified quantity of logs from land which the mill owned or on which it had acquired the right to remove timber. Customarily the mill advanced the necessary supplies from its store. At the end of the season the logger was paid for cutting whatever quantity of logs he had delivered less the value of the supplies he had used.

A mill was always utterly dependent upon its log supply. If the jobber failed to send enough logs to the saws to keep them busy throughout the season, or if the logs were of poor quality, the millowner's revenue might not be enough even to cover overhead expenses. The wording of most logging contracts was designed to prevent this catastrophe. Not only did every contract specify exactly the type and quality of the logs desired by the mill company but it usually provided that if they were not delivered on time the company could take them from the place of cutting at the jobber's expense. As a precautionary measure most companies did a portion of their own cutting each season, thus insuring that at least some of their log stock would arrive in satisfactory condition, even though it meant the outlay of more capital. If every other means of getting logs failed, it was nearly always possible to purchase cut logs from independent loggers or from lumber companies who had a surplus, but this involved a risk few mills cared to take.[23]

The sale of supplies to logging contractors was a profitable business for lumber companies owning their own stores. Lumber companies also entered into profit-splitting agreements with independent storeowners which were mutually advantageous. "If you will give us the benefit of what trade you can in the way of the jobbers on the logs," wrote a storekeeper to a mill company in 1868, "...we will at all times be ready to *divide profits* with you so far as we can." Sometimes a storekeeper and an

[23] Examples of logging contracts and bills of sale can be found in the Eau Claire Lumber Company Papers. On the need for specifying the quality of logs, see H. L. Gates to Orrin H. Ingram, October 21, 1867, and on buying logs from other mills, Knapp, Stout and Company to Ingram, June 5, 1865, both in the Ingram Papers. See also the *Northwestern Lumberman,* November 8, 1879, and Nelligan, *Life of a Lumberman,* 106.

independent logger entered into an agreement that suggests the sharecropping arrangements of the post-bellum South. The storekeeper might advance the necessary cash and supplies (sometimes at ten per cent above market value) to a logger who owned the timberland and the requisite equipment. Title to the logs as soon as they were banked became vested in the storekeeper, who could specify the time and place of delivery. Of the money received from the sale of the logs, the jobber collected only what was owing him after he had paid for the supplies and cash the storekeeper had advanced. Usually the storekeeper received also a "reasonable commission" for selling the logs and payment "for all necessary time spent in and about said business."[24]

The business of cutting logs on contract was uncertain at best, though it was sometimes very remunerative. The occasional small jobber who could pay cash for all his supplies always had an advantage over the jobber who could not; and when wages and the price of supplies were low, all contractors found themselves in a favorable position. Few of them became rich, however. Jobbers as a class, especially in the pioneering period of the industry, were constantly in debt to storekeepers and mill companies. When the log supply on the market was plentiful, contractors found capitalists reluctant to advance them supplies and money, and when logs were scarce in the market the competition was ruinous.[25]

Common interests bound the great lumbering companies together so closely that they could almost always dictate their own terms to loggers. In 1889 the large companies along the Chippewa River made a concerted effort to postpone payment of contractors until driving operations were completed in September instead of paying them in April when the camps broke up. They intended, apparently, to force the contractors to accept a ten or fifteen per cent discount if they wanted their cash at once.

[24] D. R. Moon to Orrin H. Ingram, May 16, 1865, in the Ingram Papers. Examples of share contracts, dated 1863, can be found in the Eau Claire Lumber Company Papers.

[25] *Chippewa Herald,* November 29, 1878, January 2, 1885; La Crosse *Chronicle,* February 25, 1886; *Dunn County News* (Menomonie), January 10, 1918; *Eau Claire Free Press,* October 21, 1897.

In 1893 the state legislature almost passed in the interests of the loggers' creditors a law providing for supply liens on logs and the products of logs. When a sudden change of economic conditions favored the jobbers who had already contracted for a season's work, the lumber companies were able, by virtue of their financial power, to buy back their contracts and make new ones more advantageous to themselves than the original agreements.[26] With all these handicaps log jobbing attracted ambitious men only as a stepping-stone to a higher position in the industry.

The scaling and marking of logs was done as they lay on the river shore after being banked. The measurement of logs never became an exact science. Instead of computing the actual number of cubic feet in the log the scaler, using any one of several formulas, estimated the number of board feet it ought to produce. By this method only one computation was necessary to determine how many boards of customary size could be cut from the log. Any person of moderate intelligence could use the formulas. He simply applied to the end of a log a scale rule which was calibrated to allow for waste, the tapering of the trunk, and other factors affecting the number of board feet it would produce. He could read the number of board feet directly from the rule.[27]

No scaler pretended that his measurement was absolutely accurate. The size of the saw in the mill, the skill of the sawyers, and other factors increased the inherent uncertainty of the rules. Scribner's Rule continued to be required by law in Wisconsin long after the thickness of mill saws had decreased from a half inch to the size of a knife blade. Buyers of logs and millowners profited greatly from this persistent use of an old method of measurement after improved techniques had increased the amount of finished goods that could be made from a given quantity of raw material. The logger suffered the loss, since the price of logs was always set by board measure. Occasionally a dishonest

[26] *Eau Claire Free Press,* September 10, 1889, February 9, 1893; *Wisconsin Assembly Journal,* 1893, p. 733; *Chippewa Herald,* January 2, 1885.

[27] Scribner's Rule and Doyle's Rule were the most common methods of calibration.

scaler further increased the inaccuracy of his measurements by fastening a leather pad to the end of his scale rule. By the time the diameter of merchantable logs had decreased to an average of eight inches, this device was robbing unsuspecting loggers of hundreds of dollars. The unsatisfactory techniques of log scaling led to much annoyance and frequent litigation.[28]

To facilitate the settlement of disputes over log measurement, the legislature divided the state into logging districts and empowered the governor to appoint an official scaler, called a surveyor-general or lumber inspector, for each district. It became a rule of construction of logging contracts that unless they contained stipulations to the contrary the logs must be scaled by the official inspector. In 1878 there were twelve main districts and in 1898 seventeen, and each inspector had the power to subdivide his district and appoint deputies. An inspector held office for two years and was required to give a bond of five thousand dollars, but he received no salary. His income was derived solely from the fees that lumbermen paid him for his services. Since all governors took full cognizance of the wishes of the lumbermen in a district before appointing an inspector, there is reason to believe that the office was regarded by faithful party henchmen as one more good political plum. It was not uncommon for the lumbermen in a district to divide into factions over the appointment of an inspector. These evils could produce an inefficient or dishonest inspector who would defeat the purpose of the law, but in general the courts upheld a careful inspector and in time the amount of litigation over scaling disputes decreased.[29]

[28] Bruncken, *Forests and Forestry*, 87; Matt Stapleton to Charles E. Brown, October 1, 1930, in the Brown Papers. The following excerpt from a table in the Holt Lumber Company Papers, dated November 16, 1895, shows the discrepancy. The figures represent the average number of board feet in a pine log.

Camp	*In the Woods*	*At the Mill*
2	156	150
3	103	90
5	157	121

[29] On the political aspects of the job of inspector, see John A. McRae to Orrin H. Ingram, April 29, 1872, in the Ingram Papers, and Executive Records: Lumber Inspector, 1864 ff. The *Chippewa Herald* of July 6, 1888, and the La Crosse *Chronicle* of January 31, 1889, contain items revealing inefficiency on the part of official scalers. The laws on lumber inspectors are conveniently summarized in the *Revised Statutes of Wisconsin*, 1878, ch. 84. See also *General Laws of*

The lumber inspectors also functioned as recorders of log marks. To separate his property from that of others, every lumberman getting out logs on a driving stream had his marks of ownership, which were stamped on his logs. After being recorded in the office of a lumber inspector these marks became his exclusive property. Anyone using an unrecorded mark or a mark recorded by another was subject to fine, half of which went to the prosecuting lumberman. From time to time legislation was passed requiring the lumbermen in specified districts to re-register their marks, failing which the mark would be considered abandoned. Each lumberman selected at least two marks, one to be placed on the bark near the end of the log and the other to be stamped on the end. The bark mark, which usually consisted of straight lines, was cut by the swamper with a few deft strokes of an axe before the log reached the skidway. The other, applied while the logs stood on the skidways, was stamped on the end with a hammer bearing an embossed design on its head. In 1868 the superintendent of booms at Chippewa Falls requested loggers to place their bark marks near the middle of the log, where they would be less likely to be defaced, and to place from five to eight end stamps near the sap line.[30]

Banking and marking completed the winter job of the camp crew. The logs were now ready for the trip to the mill.

Wisconsin, 1864, ch. 167; 1868, ch. 129; 1869, ch. 74; 1870, ch. 90; 1871, ch. 70; 1875, ch. 242; 1876, ch. 254; 1881, ch. 225; 1882, ch. 193; 1885, ch. 112; 1889, ch. 387; 1895, ch. 54; and 1897, ch. 74.

[30] *Revised Statutes,* 1878, sec. 1738; *General Laws,* 1883, ch. 267; 1889, ch. 441; *Eau Claire Free Press,* October 4, 1883, May 9, 1889; *Chippewa Herald,* May 3, 1889; *Chippewa Union and Times,* December 26, 1868. The Brown Papers contain a long list of log marks. For identification of the log marks used as decorations in the chapter heads of this book, see Lumber Inspector, District 14, Register of Log Marks, 1882–1894, in the possession of the State Historical Society of Wisconsin.

4. From Forest to Mill

WHEN the rays of the spring sun had broken up the ice in the rivers, it was imperative to get the banked logs started on their way to the mills, for logs left on the rollways were subject to damage from fire, worms, timber-borers, and sap stains. At the end of a single season in the woods the logs would be perforated by worms to the depth of an inch, and the wood discolored by sap. This damage could be minimized by peeling the bark from the logs and spreading them singly on the ground or by stacking them properly on skids in a dry place, but even with these precautions the damage was likely to amount to twenty per cent by the end of a single year. Logs left in the woods for two years were all but ruined.[1]

Driving conditions, which varied from one stream to another, greatly influenced the pattern of urban development in the several lumbering districts. Along the Wisconsin River, on which driving was difficult, the mills were built as close as possible to the area of cutting. Where driving was easier, as on the Wolf, Black, Chippewa, and St. Croix rivers, the manufacture of lumber became more centralized. Thus Oshkosh, La Crosse, Eau Claire, Chippewa Falls, and Stillwater developed into important milling towns in their respective districts. Logs were

[1] *Northwestern Lumberman,* March 16, 1878; *La Crosse Tribune and Leader-Press,* March 13, 1938.

brought to Eau Claire from areas within fifty miles of Lake Superior. Other cities, such as Fond du Lac and Marinette, received their logs in rafts towed from the driving terminus. After 1870 Wisconsin logs helped to develop certain upper Mississippi sites into important log-rafting and sawing centers.[2]

Driving was the most picturesque operation of the lumber industry.[3] As the logs broke from the rollways they joined the chunks of broken ice in the river with a cloud of spray and a tremendous splash and gathered momentum in the current. Down the stream they flowed unchecked, twisting around bends, leaping over falls and through the sluiceways of dams, smashing over treacherous rapids, until their volume and pressure became too great for the width of the streambed, and they stopped, jammed into a confused mass, past which the water continued to flow only with difficulty. It was as if some gigantic hand had cast a stack of jackstraws carelessly into the waters with their ends protruding in every direction.

The rivermen, "picturesque and heroic figures in truth," rode the backs of the driven logs, their bodies heaving and swaying to the surging vagaries of their unstable supports. They were a set of "hard-living, hard-drinking, hard-fighting, blasphemous pioneers," picked from the camp crews for their stamina, agility, and strength.[4] Their task was to keep the unruly logs always moving and to gather up the recalcitrant ones that became stranded in the eddies and backwaters of the stream. Adventure was their daily companion until their charges were safely boomed near the mill.

When the logging camps broke up in the spring, most of the lumberjacks set a course for the nearest town, where in a few

[2] John M. Holley, "Waterways and Lumber Interests of Western Wisconsin," in *Proceedings of the State Historical Society of Wisconsin,* 1906, pp. 208–215; *Eau Claire Free Press,* November 1, 1883; Dopp, "Geographical Influences in the Development of Wisconsin," in the *Bulletin of the American Geographical Society,* 45:739; Nelligan, *Life of a Lumberman,* 112–113. See also below, Chapter 9, for a discussion of the development of sawmilling on the Mississippi.

[3] The following description is based on Bruncken, *Forests and Forestry,* 85–86; Nelligan, *Life of a Lumberman,* 134 ff.; *Eau Claire Free Press,* May 21, 1889, June 15, 1893; letter of Augustus Cole to Holt and Balcom, June 16, 1887, in the Holt Lumber Company Papers; and Lincoln, Lumbering in Early Days (MS.), 5–6.

[4] Nelligan, *Life of a Lumberman,* 134.

riotous days many of them spent a winter's wages. A few remained behind in the woods during the month between the breakup of camp and the beginning of the drive to repair as best they could the trails along the river courses, to blast out rocks in the streambeds that might start a jam, and to make the course of the logs as easy as possible. Before the drive began they were joined by reinforcements, and they prepared for action. The flannel and woolen clothing of winter give way to overalls, a lighter garment when wet, and quicker to dry. Shoepacs and rubbers, the standard winter footwear, were replaced by heavy calked boots to make traveling on the logs easier. Pike poles and peaveys were taken out and inspected, for during a drive a man's very life depended on his tools.

The original tools of the rivermen were the pike pole and the jam pike. The pike pole was about sixteen feet long, the jam pike only five feet. Both had a sharp steel point on the end. By driving the jam pike into one log and standing on another the riverman could ride the drive in comparative comfort. The pike pole was essentially a tool to manipulate logs into position, its length enabling a man to reach a log caught on the shore, and its point enabling him to maneuver a log in the water. Occasionally a cant hook, a five-foot hardwood stick with an adjustable hook near one end, was used on the drive, though it was most commonly employed in the camp and in the mill to turn logs on the skidways or on the mill carriage. In 1858 an ingenious Maine blacksmith conceived the idea of combining the cant hook and the jam pike. The result was the peavey, an extremely versatile tool which could be used as lever, pike, or gripping instrument.[5]

When the drive got under way the crew was divided into two or three parts, depending upon the size of the stream. One group kept in the van, riding on the logs or in a boat, to prevent jams from forming and break them if they did form, to ease the logs over dams, and in general to see that the mass kept moving. The

[5] The word "cant," meaning a sudden thrust resulting in a change of direction, is a nautical term and was probably brought to Wisconsin by Yankees who had worked in shipyards. In the mill a cant was a log whose four sides had been sawed off. The peavey was named after its inventor. The principal difference between the cant hook and the peavey was the point in the end of the latter. In Wisconsin the terms appear to have been used interchangeably at

remainder, one group on each side of the river, brought up the rear to pick up the logs that became stranded. These hard-working rivermen consumed four or five heavy meals during their long workday unless the exigencies of the drive precluded regular meals. On long drives the cook's wanigan followed the rear crews, moving ahead through the logs whenever possible so as to have a steaming meal ready at the proper time. When the drive moved too fast for the cook, the men were forced to eat the cold canned food they carried in haversacks or "nosebags." At night, if it could be arranged, they slept in tents supplied by the driving company; more often than not they slept in the open.

It was a laborious life, fraught with danger. Up at three o'clock for breakfast, the driving crew walked for miles along the bank or rode dangerously in the midst of floating logs. There was no day when they were not in the water, handspiking logs off rocks or sand bars, and trying to keep the logs from jamming. Their clothes were never dry until they hung them before the fire in the evening. In all sorts of weather they performed their exhausting series of tasks, and when night came they arrived in camp weak with fatigue. Danger was a commonplace until the booms were reached. "The jam crackers are sort of daredevil, skillful fellers, who won't work anyplace where there ain't danger," wrote a member of a driving crew in 1889. If a driver was so unfortunate as to be caught in the rush of logs from a rollway, he was crushed beyond recognition. Even the most skillful riverman fell from his perch occasionally, and if he was caught under a mass of logs or in a whirlpool, his comrades were helpless to save him; he drowned before their very eyes. Riding the logs, steering them over the sluiceways of dams, breaking jams, all of it was hazardous in the extreme and often brought tragedy, which the riverman learned to accept callously. For all this he received about fifty cents a day in the 1840's and about two dollars and a half fifty years later.[6]

times. Usher, *Wisconsin,* 1:171; Fernow, "American Lumber," in Depew, *American Commerce,* 1:201; Otis W. Terpening to Charles E. Brown, March 12, 1931, in the Brown Papers.

[6] *Eau Claire Free Press,* May 21, 1889, June 15, 1893; Lincoln, Lumbering in Early Days (MS.), 5–6; Stephen R. Bentley, An Account of Wisconsin River

The most dangerous of all the driving operations was the breaking of a large log jam. A narrowing of the streambed, a rocky section, or a stretch of shallow water was likely to start a jam. At the height of the driving season, in April or May, when logs were being floated by all the operators, a jam, once started, quickly piled up a stupendous amount of timber. When in 1869 a boom above Chippewa Falls broke, a jam formed in the rapids below the city which ultimately heaped some hundred and fifty million board feet of logs into a tangled mass extending fifteen miles upriver and thirty feet into the air at some points. Another jam which formed at Eagle Rapids in 1874 contained a hundred million board feet. Two miles above Menomonie a gorge formed in 1876 which was three miles long, nearly thirty feet high, and an eighth of a mile wide. In 1879 a solid jam nearly fifteen logs high and thirteen miles long accumulated on the Jump River and another at Wausau which was nearly four miles long. The mass of matted logs that became lodged on the Chippewa in 1880 comprised ninety million board feet. In 1886 more than two hundred men and two steamboats struggled for six weeks to break a jam of a hundred and fifty million board feet on the St. Croix.[7] The cost of breaking one of these jams ran into thousands of dollars. Before the day of cooperative driving companies a group of log owners would organize a committee to supervise the work of breaking a jam and prorate the cost in proportion to the number of logs each had in the stream.[8]

If the log mass was a comparatively small one, it might sometimes be demolished by a large crew of men working with pea-

Lumber Rafting Days, p. 4, manuscript dated September 30, 1911, in the possession of the State Historical Society of Wisconsin; Stephenson, *Recollections*, 88. In Stephenson's view, "subjecting them to hardships which, from the latter day point of view, might have been considered inhumane, seemed to do more good than harm." The most dangerous work was done by those who "rode the drive," but the follow-up crews received a higher rate of pay because their work was harder.

[7] *Eau Claire Free Press*, May 6, 1869, November 29, 1900; La Crosse *Republican and Leader*, May 23, 1874, June 10, 1876; *Chippewa Herald*, May 23, 1879, April 30, 1880; *Northwestern Lumberman*, May 31, 1879; Folsom, *Fifty Years in the Northwest*, 708.

[8] *Eau Claire Free Press*, June 28, 1888; Pound, Halbert & Co. to Ingram and Kennedy, October 25, 1869, in the Ingram Papers.

veys in midstream. If the jam was a large one, however, this was too hazardous a procedure. Instead, teams of horses, aided by men with peaveys, would pull out the logs along the shore on both sides of the jam until a section was detached from the banks and left hanging like an apron. When this apron became large enough, the pull of the current would finally break it off, and it would float away from the main mass. The process was then repeated until the entire jam had been dispersed. On a particularly difficult job dynamite was sometimes used; it was efficient and comparatively safe, but it was likely to damage too many logs to be acceptable as a routine weapon.[9]

The breaking of a jam was an event of great importance to a town, and the spectacle of the final operations often attracted a cheering audience:

> Men, women, and children gather on the bank to see the sight; the drivers, in red flannel shirts, dripping wet, shout and labor; the oxen strain and pull; here a log starts; another one is hauled out; slowly the head of the jam gives way; shout upon shout rends the air, and amid the cheers of those looking on and the waving of hats, bonnets and aprons, the jam breaks in front and moves on like an army, to "halt," perhaps within half a mile where the same labor must be repeated.

In 1882 an Oshkosh newsman described vividly the tense drama he had witnessed when a quiet, twenty-year-old riverman cut the last log holding up a large jam:

> He stripped everything, save his drawers. A strong rope was placed under his arms, and a gang of smart young fellows held the end. The man shook hands with his comrades, and quietly walked out on the logs, ax in hand .... The man was quietly walking to what might be his death. At any moment the jam might break of its own accord, and also, if he cut the log, unless he instantly got out of the way, he would be crushed by the falling timber.
>
> There was a dead silence while the keen ax was dropped with force and skill on the pine log. Now the notch was near half through

[9] Nelligan, *Life of a Lumberman,* 136; *La Crosse Tribune and Leader-Press,* March 20, 1938. Some rivermen believed that a single log, the "key log," held the entire jam, and that when it was found and removed the gorge would disappear. It seems likely, however, that no large jam was created by a single

the log; one or two more blows and a crack was heard. The men got in all the slack of the rope that held the axman; one more blow and there was a crash like thunder and down came the wall, to all appearances on the axman.

Like many others, I rushed to help the poor fellow, but to my great joy I saw him safe on the bank, certainly sadly bruised and bleeding from sundry wounds, but safe.[10]

Lack of rain during the driving season fostered the formation of jams. On most of the driving streams in the state a dry season meant that "sand bars and rocks would be more plentiful than water"; a ten-day rain, on the other hand, might raise the water level as much as two or three feet and enable the driving crews to get to the mills logs that had been hopelessly stranded. Transportation from the mill to the lumber market also depended upon the level of the rivers. Thus rain or the lack of it meant life or death to a lumbering community.

Inhabitants of a logging and lumbering region know full well what a "dry season" means. It points to empty pocketbooks, idle days, and a Tantalus of weeks and months, where owners are compelled to sit upon the banks of evaporating streams, and see their only source of revenue or fortune, lying high and dry upon the burning sands, oozing pitch from countless cracks, or becoming food for worms. It means notes coming due, and mid-summer payments to make—and it means the very quintessence of annoyance, a dull, heavy quiet, instead of the exciting rush of machinery, the cheering halloo of the gangs of "drivers," the shouts of the raft-runners, and the grand turmoil incident to this species of existence.

With the introduction of the logging railroads lumbermen were less dependent on the height of the rivers. Meanwhile it was possible to bring about artificially a rise in the water level by a system of flooding dams on the upper reaches of the driving streams.[11] These dams, constructed of heavy timbers, gravel, and

log, although it is not impossible that several logs in strategic positions could hold up an entire drive.

[10] *La Crosse Weekly Democrat,* May 29, 1865; Oshkosh *Northwestern,* May 18, 1882.

[11] *Chippewa Falls Democrat,* March 23, 1872, from which the quoted passage is taken; *Chippewa Herald,* July 13, 1872; La Crosse *Republican and Leader,* August 7, 1875; *Eau Claire Free Press,* August 4, 1898; *Northwestern Lumberman,*

earth, continued to be extensively used in areas where railroad construction lagged.

In the pioneer days of the industry each owner put his logs into the stream at his own discretion and allowed them to mingle with those of others. The result was utter confusion, and bitter conflicts were regular vernal events. One driving crew might be delayed by a slower crew ahead. A flooding dam built by one logging company for its own purposes might hamper every other driving crew on the stream by changing the water level at the wrong times. When the logs did arrive at the mill, they were mixed almost inextricably with those of other mills. Each mill found it necessary to construct sorting works where its own logs could be segregated and the others passed on down the stream. These private sorting works, however, often proved inadequate to handle all the logs that came along. Downriver mills were subjected to the annoyance of having their logs retained by works upstream, and often during the rush of a heavy driving season a mill would be forced to let some of its own logs pass on. To meet this situation, practical lumbermen developed a system of exchanging logs; but equitable log exchange, calling as it did for a great deal of laborious bookkeeping, never worked to their satisfaction and did not prove to be the answer to the problem.[12]

With the tremendous growth of the industry during and after the Civil War it was recognized that some central organization of the lumbermen on a given river was necessary, not only to handle the logs on the drive but also to build booms, or obstructions, where the logs could be held and distributed among their rightful owners. Such an organization could perform other services as well. It could construct and maintain flooding dams, remove obstructions to safe driving, and gather up the logs that became stranded on the riverbanks.[13]

August 25, 1877; Oshkosh *Weekly Northwestern,* May 17, 1888; W. A. Holt to Holt and Balcom, May 20, 1887, in the Holt Lumber Company Papers.

[12] Oshkosh *Weekly Northwestern,* May 17, 1888; Nelligan, *Life of a Lumberman,* 90, 115–116, 159 ff.; Log Survey, July 27, 1862, in the Ingram Papers. For an agreement between two companies to use each other's dams, see W. A. Holt to Holt and Balcom, June 20, 1887, in the Holt Lumber Company Papers.

[13] Stephenson, *Recollections,* 162; Oshkosh *Weekly Northwestern,* May 17, 1888.

WHi(W6)11411

Chippewa River, Wisconsin, 1869. Log jam of about 150 million board feet of logs, stopped at the piers of Pound, Halbert & Co., which stood in about forty feet of water. In some places the logs were twenty feet above the water.

WHi(X21)191

Peshtigo River, Wisconsin, probably 1890 – 1910. Log drivers working logs through a pond sorting area. This operation separated logs belonging to different owners and separated logs destined for different uses.

Even before 1860 the lumbermen on a few logging streams instituted booming and improvement companies on a small scale, using the New England industry as a model. At every session from 1850 until the disappearance of large-scale lumbering the legislature was besieged with demands for charters empowering individuals or companies to build dams, construct booms, or improve rivers. Between 1850 and 1873 the legislature passed more than a hundred boom acts and more than two hundred dam acts, including amendments to bills previously passed. As early as 1850 the general form of improvement bills was laid out by Moses M. Strong of Mineral Point, who envisioned a project for developing Grand Rapids on the Wisconsin River and set about drawing up a bill that would permit the backers of the scheme to collect a toll from everyone using the river. Such a toll was of course a legitimate and necessary charge, but the public was disposed to assume that the company's profits were excessive, and there was much complaint from those who had to pay it. In 1883, for example, certain loggers on the Chippewa took exception to the number of bills passed which enabled "John Smith and Peter Jones" to build dams "so that the pockets of Jones and Smith aforesaid may become plethoric by means of tolls assessed."[14]

By 1870 every logging district had its booming and improvement company. The Wolf River Boom Company was organized in 1857 and served the lumbermen of its district throughout the logging era despite several reorganizations and a bankruptcy action in 1871. In 1876 the loggers of the Wolf River, the Red River, West Branch, and Shawano made arrangements to consolidate their drives. Three years later the Wolf River Boom Company was maintaining eight miles of receiving booms twenty miles from Oshkosh and sorting works in a bay of Lake

[14] Wood, *Lumbering in Maine,* 232; letters to Moses M. Strong from John Werner, Jr., January 18, 1850, and from George M. Strong, July 12, 1856, in the Strong Lumber Papers; Oconto River Improvement Co. to T. J. Cunningham, January 15, 1894, and Northwestern Improvement Co. to R. McMillan & Co., June 11, 17, 1889, in the Holt Lumber Company Papers; *A Synoptical Index of the General and Private and Local Laws of Wisconsin from the Organization of the Territory to 1873 Inclusive* (Madison, 1873), 41–43, 108–112; *Chippewa Herald,* April 6, 1883.

Poygan. Forty dams built on the upper river also aided the log drivers.[15]

On the Wisconsin River the Little Bull Falls Boom Company was organized in 1852 and four years later the Wisconsin River Boom Company. By 1879 Wausau had booms large enough to hold thirty million board feet safely, yet the next year the city was being petitioned to take ten thousand dollars' worth of stock in the boom company so that its capacity could be enlarged. In 1882 a meeting at Wausau resulted in the formation of a driving association for the Wisconsin River and its tributaries.[16]

The first driving and booming company serving all Black River lumbermen was formed in 1854, superseding the local company that had been started at Black River Falls a year earlier. In 1864 the Black River Improvement Company was organized under a franchise granting it almost unlimited powers to handle loose logs along the entire length of the river. By 1872 another organization, an unincorporated log-driving association, was also operating successfully, and in 1879 a flooding dam association was organized to improve driving conditions.[17]

The Chippewa River Booming Company was organized in 1855, probably with the aid of capital from New York. For many years thereafter the quarrels of lumbermen at Eau Claire and Chippewa Falls and the opposition of all millowners to the log-driving projects of Mississippi River lumbermen hampered river improvement and the advancement of booming projects on the Chippewa. Both conflicts were resolved by 1880, and a pool known as the Chippewa River Improvement and Log Driving Company was established to control the movements of loose logs all along the river. Eau Claire, the largest milling town on the

[15] Hotchkiss, *Lumber and Forest Industry*, 390; Oshkosh *Northwestern*, February 2, March 16, 1871, June 22, 1876; *Northwestern Lumberman*, April 5, 1879; Schafer, *Winnebago-Horicon Basin*, 288–289.

[16] *Private and Local Laws of Wisconsin*, 1852, ch. 323; 1856, ch. 76; *Northwestern Lumberman*, March 29, 1879; La Crosse *Chronicle*, January 29, 1880; *Chippewa Herald*, January 12, 1883.

[17] *Private and Local Laws*, 1853, ch. 213; 1854, ch. 52; *La Crosse Tribune and Leader-Press*, March 6, 1938; Usher, *Wisconsin*, 1:184; La Crosse *Republican and Leader*, March 2, 1872, February 8, 1873, February 22, 1879.

Chippewa, developed a huge storage pond and sorting works at the Dells above the city. The interests of Mississippi River mill-owners were served by the great Mississippi River Logging Company.[18]

The fact that the St. Croix River formed part of the boundary between Wisconsin and Minnesota produced constant friction between the two states over their respective rights to regulate log driving and inspection on the river. In 1851 Minnesota Territory and Wisconsin each chartered a boom company to serve lumbermen on the St. Croix, and both companies continued to operate until 1858, when the Wisconsin charter was repealed. In general, the lumbermen along the St. Croix came to look to Minnesota for protection, although Wisconsin continued to issue booming charters and the courts declared that the two states had concurrent jurisdiction. The Menominee River Boom Company, operating on the Wisconsin-Michigan boundary, was less hampered by interstate disputes. Between 1868 and 1894 it handled nearly eight billion board feet of logs.[19]

No boom nor improvement company could at any time satisfy the needs of every lumberman on a driving stream. Complaints from dissatisfied drivers were constantly coming into the offices of such companies—complaints that "splashes" from flooding dams did not come at the right time, that logs were being held up too long in the sorting works, that too many logs were stranded during the drive. These criticisms were sometimes deserved. One company, it was charged, failed to move forty million feet of logs belonging to the Mississippi River Logging Company which in the years 1890–95 became stranded in the flowage of the Dells of the Chippewa River. Disputes between individual loggers also continued, but the state developed legal means for their settlement.[20]

[18] *Private and Local Laws,* 1855, ch. 331; Ely, Kelly, and Vail to Ingram and Kennedy, January 29, 1866, in the Ingram Papers; *Eau Claire Free Press,* November 1, 1883; Hotchkiss, *Lumber and Forest Industry,* 635.

[19] Folsom, *Fifty Years in the Northwest,* 696; *Private and Local Laws,* 1851, ch. 351; 1858, ch. 191; Hugh H. Price to Orrin H. Ingram, February 27, 1889, in the Ingram Papers; *Annual Report of the Menominee River Boom Company* (Marinette, 1894), cover.

[20] Letters to Orrin H. Ingram from Joseph G. Thorp, August 16, 1875, and

To handle efficiently the thousands of logs that came down the rivers of the state each year, the boom companies constructed extensive works in the stream, often stretching for miles. The booms were of two types, jam booms and sheer booms. The jam boom was designed to stop the logs in their passage and retain them until they could be handled by the mills. It was made of strong timbers stretched rigidly across the channel and buttressed with piles driven into the river bottom or with cribs filled with stone. It was usually located at the foot of an island, which became one side of a retaining pocket.[21]

In connection with such a jam boom, sorting works were constructed. At the lower end of the pocket formed by the boom an opening permitted the logs to enter a narrow passage flanked by gaps that led to the booms of individual owners or to a place where log rafts were put together. Over each of these gaps a workman, armed with a pike pole, stood on an elevated platform to steer the logs bearing the appropriate marks into his gap. Logs destined for mills farther down the river were either made into rafts at the sorting works or allowed to pass on to another sorting place. This method of dividing logs could be very efficient. The sorting works of the Black River Improvement Company, for example, could handle a million board feet of logs a day. During rush periods it was common to sort at night, especially after electric lighting came into use.[22]

A sheer boom was a device to divert moving logs from the main channel into a jam boom or other works. Its construction presented many problems to the lumber industry. Obviously any obstruction designed to shunt moving objects into a circumscribed portion of a stream must stretch all the way across the channel. It must be stiff and strong, and wide enough for a man to walk upon; the side facing upstream must be smooth, so

from Frederick Weyerhaeuser, July 31, 1895, in the Ingram Papers; La Crosse *Chronicle*, January 17, 1889; William J. Hagenah, *The Law of Labor in Wisconsin* (Madison, 1908), 126; *Statutes of Wisconsin*, 1898, sec. 3337.

[21] *Eau Claire Free Press*, September 24, 1874.

[22] *La Crosse Tribune and Leader-Press*, March 30, 1938; *La Crosse Weekly Democrat*, May 29, 1865; La Crosse *Republican and Leader*, May 4, 1878; Hotchkiss, *Lumber and Forest Industry*, 142.

that logs would not lodge against it. At the same time, because the law forbade any hindrance to traffic up and down a navigable stream, it must be instantly removable from the main channel. Early lumbermen designed a clumsy method of removing their sheer booms from the channel with windlasses, but it remained for James Allan and Levi Pond of Eau Claire to discover in 1861 how the force of the current could be utilized both to hold the boom across the channel and to remove it when a boat or raft passed by. Their device, one end of which was anchored to shore, consisted of a rigid chain of timbers upon which was fastened a series of collapsible rudders, or fins. When the rudders were extended at an angle of thirty degrees or more on the lower side of the boom, the current pushed against them with such force that the boom was held obliquely across the channel. One man on the bank could easily pull the rudders into a position parallel with the boom, whereupon the current would immediately push the boom aside to permit the passage of river craft. So adequately did the fin boom meet the need for which it had been devised that it was soon in use on every navigable logging stream in the country.[23]

Despite the best efforts of the driving and booming companies, many of the logs that had been taken from the woods failed to reach the mill. Part of the discrepancy between the logger's and the mill's scaling can be attributed to inaccuracy at the rollways, but some of it represented actual loss of logs, resulting from their being stranded in stagnant bayous and backwaters or on the banks of low channels. Sometimes this loss was considerable. In 1855 thirty million feet lay "high and dry" in the swamps of the Mississippi River six miles above the mouth of the Black River, and in 1875 an equal amount washed ashore on the Black. Even in good driving years and with well-organized driving companies the loss was considerable. In 1886 a large operator on the Chippewa declared that his company must allow for at least ten per cent loss every year. During years of drought the percentage increased. For at least five years before

[23] *Eau Claire Free Press,* September 24, 1874; Merk, *Economic History of Wisconsin during the Civil War Decade,* 67–69.

1873 the annual loss on the Chippewa amounted to thirty per cent, and in 1883 it soared to forty per cent.[24]

After 1856 the law provided that if a log owner failed to remove his stranded logs from the shore within eighteen months, the owner of the land upon which they lay might, after advertising, seize and sell them at auction for twice as much as the costs of removal and damage to the land. Accordingly the log owners made every effort to arrange for the restoration of the logs to the driving stream as soon as possible. Before the day of booming companies it had been customary for log owners to sell their stranded logs to anyone who wished to gather them up, or to let a contract in advance for "sacking in," as such salvage work was called. If heavy spring floods had caused an unusually large number of logs to be washed ashore, the lumbermen along a driving stream might appoint a committee to supervise the sacking in for all firms. The booming companies, however, usually handled such work by letting contracts for gathering up the logs during the summer and distributing them to their owners. They also appointed watchmen to patrol the river and prevent the theft of stranded logs, for the sloughs and bayous were "most Excellent Places for Jayhawking." This log-salvage work often damaged farms and crops along driving streams. The law provided for the payment of damages, but many lumber companies were reluctant to make such restitution, with the result that relations between farmers and loggers were strained much of the time.[25]

Another driving problem that the booming company handled was the disposal of prize logs, strays, and scrabble. The law defined a prize log as one without an owner's mark which had not been claimed within a year after the general drive. Such logs

[24] *La Crosse Independent Republican,* July 11, 1855; La Crosse *Republican and Leader,* July 17, 1875; *Chippewa Herald,* February 8, 1873; *Eau Claire Free Press,* July 5, 1883, January 21, 1886.

[25] *General Laws,* 1856, ch. 48, sec. 1, 4; 1858, ch. 31, sec. 2, 3; 1881, ch. 252, sec. 1; *Revised Statutes,* 1858, ch. 42, sec. 1, 4; *Statutes of Wisconsin,* 1898, sec. 1600; F. B. Andries to Jackson and Ingram, January 14, 1867, N. Bowman to O. H. Ingram, January 7, 1867, Philipp Schechel to Dole, Ingram, and Kennedy, January 17, 1867, minutes of a meeting of lumbermen, dated June 23, 1880, and J. B. Randall to Ingram and Kennedy, April 13, 1865, in the Ingram Papers; *La Crosse Tribune and Leader-Press,* March 13, 1938.

were prorated among the companies in a subdistrict on the basis of the number run in the general drive. The strays were marked logs that had been misdirected at the sorting works. These were caught in booms farther down the river and the owners were paid for them at the end of the season. Loose logs, marked or unmarked, that had for some reason escaped being rafted were called "scrabble." Most boom companies maintained scrabble booms near the end of their driving streams.[26]

From the early days of the industry the practice of rafting logs rather than driving them loose had been common. It was superseded only where security was a less important consideration than economical driving. Where loose driving entailed inordinate loss, as it did on the Mississippi or a large lake, the logs had to be kept under control. The booming companies therefore assumed the additional function of preparing log rafts for mills at a distance from driving streams. Many individuals, too, conducted rafting businesses.

In 1843, a year after a group of Mormons had demonstrated the feasibility of rafting logs from the Black River to Nauvoo, Illinois, the advantages of the practice became apparent to lumbermen along the Mississippi. In that year a boom on the St. Croix broke and allowed all the logs of a certain lumber company to escape with the current. Unable to return them to its mill, the company gathered as many as it could at Stillwater, formed them into four rafts of a half million board feet each, and sold them at St. Louis for a fair price. From this beginning the business of log rafting grew as mills sprang up along the Mississippi. Steam towboats appeared on the St. Croix as early as 1848 and on Lake Winnebago during the fifties, but it was not until the late sixties that they became important as motive power for log rafts on the Mississippi. After 1870 it was common practice to raft logs on the rivers of all the lumbering districts and on the Great Lakes.[27]

[26] *Revised Statutes,* 1878, sec. 1740; *La Crosse Tribune and Leader-Press,* March 13, 1938.

[27] Durant, "Lumbering and Steamboating," in *Minnesota Historical Collections,* vol. 10, pt. 2, pp. 649, 665; Holley, "Waterways and Lumber Interests," in *Proceedings of the State Historical Society of Wisconsin,* 1906, p. 212; Usher, *Wis-*

Several methods of constructing log rafts were devised. In the early days a raft that was to travel only a short distance was held together by a rope fastened with half hitches to plugs driven into holes in the logs. If the trip was a longer one, the raft was locked together by poles laid across its width and wedged into holes in the logs. Both these methods were superseded after 1870 by the more economical brail system. By this technique the logs to be rafted were enclosed in a hollow frame, held together by binding chains and rope braces, which eliminated the damage done by boring holes in the logs. A standard brail, as constructed for driving on the Mississippi, was six hundred feet long and forty-five feet wide. Six brails fastened together made up a Mississippi raft, which might contain as much as a million feet of logs and cover an area of three acres.[28]

The lumbermen agitated constantly for publicly financed river improvements. Wisconsin lumbermen were well represented at a series of "river conventions" called after the Civil War by transportation groups to promote a long-term program of improvements along the Mississippi from New Orleans to St. Paul. Individual millowners also exerted pressure on their representatives in both federal and state legislatures. Some of them wrote letters describing their needs in great detail, even in a tone of instruction rather than suggestion or advice. Not infrequently lumbermen fully cognizant of the needs of the industry were themselves state legislators or members of Congress.[29]

Influenced by such pressure groups and yielding to the spell of the lush post-war economic expansion, Congress after 1865 appropriated increasing sums for Wisconsin lakes and rivers. Government engineers repeatedly surveyed logging streams and supervised the deepening of harbors and channels, the removal of obstructions, and the development of flowage control. The

consin, 1:171; Schafer, *Winnebago-Horicon Basin,* 270; *Chippewa Herald,* August 21, 1891.

[28] Durant, "Lumbering and Steamboating," in *Minnesota Historical Collections,* vol. 10, pt. 2, p. 665; Blair, *Raft Pilot's Log,* 51; Emil Heintz, "Rafting on Black River," in *La Crosse County Historical Sketches,* Series 3: *The Lumber Industry,* 36–39; Harry Dyer, Log Rafting on the St. Croix River, manuscript dated August, 1912, in the Brown Papers; *Chippewa Herald,* June 5, 1874.

[29] *Eau Claire Free Press,* October 18, 1877; *Chippewa Herald,* October 24, 1879,

most grandiose scheme of its time was that proposed in 1879 by Thaddeus C. Pound, a Chippewa Falls lumberman who represented his district in Congress. His plan called for the creation of artificial reservoirs at the headwaters of the Mississippi and its tributaries so that the water level could be kept fairly constant the year around. Although the project was enacted into law, the difficulties involved in securing flowage rights and the reluctance of certain lumbermen to support the plan because they feared it would aid their natural enemies, the steamboat interests, delayed the construction of the reservoirs, but by 1885 some progress had been made. The project was finally abandoned when the reservoirs failed to work as had been expected.[30]

In any event, lumbermen did not regard the river-improvement work of the government as an unmixed blessing. They were always apprehensive lest the government construe their dams, piers, and booms as obstructions to navigation and order their removal, and they sought constantly to ascertain the plans of the secretary of war regarding improvements. They knew very well that what constituted an improvement to lumbermen was often detrimental to steamboat transportation. The running of loose logs in great numbers, the regulation of stream flow at will by private flooding dams, and the dumping of sawdust and other mill waste into navigable streams all hampered steamboat navigation. The opposition of steamboat companies to these common practices of the lumber industry bore fruit in legislation forbidding the dumping of sawdust into rivers, and in efforts to limit the running of loose logs on streams capable of carrying other traffic. In 1899 the federal rivers and harbors bill was so amended as to prohibit the floating of logs, timber, and rafts on navigable rivers. Minnesota and Wisconsin congressmen succeeded in exempting from this provision the most

August 28, September 11, 1885; William F. Price to Orrin H. Ingram, January 14, 1885, in the Ingram Papers.

[30] *Eau Claire Free Press,* April 4, 1874; *Chippewa Herald,* September 6, 1878, October 31, 1879, April 16, 1880, February 4, April 29, August 12, 1881, May 4, 1883, September 25, December 11, 1885, July 26, 1895. In 1907 the state authorized a private corporation to experiment with a similar plan to control flowage on the Wisconsin River. *Report of the State Forester of Wisconsin, 1907–08,* pp. 85–90.

important logging streams tributary to the Mississippi, but the lumber industry was forced to accept a provision allowing the secretary of war to make regulations governing navigation on the exempted streams. Despite their basic antagonism, steamboat and lumber interests learned to cooperate when they were faced with a common danger, such as the construction of railroad bridges that were likely to be detrimental to navigation.[31]

The log-driving interests had other enemies, among whom were the riparian owners who suffered damages from the practices of the log drivers. Although such differences were often amicably adjusted, little good feeling was lost between the two groups. Artificial flooding by private dams was especially resented by the landowners, who complained bitterly of the damage done to buildings, fences, soil, and cattle along the stream bank. At times the landowners became so irate that the owners of flooding dams had to provide guards for their property. In 1884 a measure making improvement companies liable for such damages was defeated by only one vote in the Assembly. It was in the hope of avoiding responsibility for damages caused by the flooding dams that many lumbermen supported Pound's artificial reservoir plan. As a backer of the project wrote in 1885, "The U. S. government can do things that individuals cannot, and hence the only way to continue flooding is to let the government do it. There is no question but what the government would use the dams in the interests of the lumbermen, and the operators on the river."[32]

Among the enemies of the log drivers were the "river pirates" who stole logs on their way to the mill. During the seventies it was common practice for unscrupulous men living along a driving stream to cut shingle bolts during the night from loose logs

[31] M. D. Bartlett to Ingram and Kennedy, March 14, 1872, in the Ingram Papers; *Sources of the Mississippi, Saint Croix, Chippewa, and Wisconsin Rivers* (45 Congress, 3 session, House Executive Document no. 54, serial 1858, Washington, 1879); *Loose Logs, Steamboats, and Rafts on Certain Rivers and Streams* (56 Congress, 1 session, Senate Report no. 1035, serial 3890, Washington, 1900); *United States Statutes at Large*, 31:172; *Chippewa Herald*, January 23, June 18, 1880, May 7, 1886, March 2, 23, 1900. See also below, Chapter 9.

[32] George C. Ginty to Orrin H. Ingram, January 3, 1885, in the Ingram Papers; *Chippewa Herald*, March 6, 1885.

floating past their homes. River police had to keep constant watch over logs that became stranded on the banks. Even log rafts were not safe from thieves. In 1900 a gang of twenty-two men succeeded in taking a half-raft from the Mississippi near Dresbach and selling it at La Crosse, thus starting a wave of similar crimes. Some millowners employed watchers to insure that mills farther upstream did not use any of their logs. Special legislation provided penalties for sawing the logs of others without their consent, but millowners hesitated to proceed singly against such offenders lest the river pirates cut the booms of their company and cause still greater losses.[33]

[33] L. H. Cushing to Ingram and Kennedy, June 25, 1874, S. W. Thomas to Orrin H. Ingram, April 22, 1872, and W. P. Tearse to Empire Lumber Co., May 2, 1902, in the Ingram Papers; *Eau Claire Free Press,* August 16, September 27, 1900; *General Laws,* 1855, ch. 60, sec. 1; 1856, ch. 48, sec. 4; 1864, ch. 263, sec. 1; 1866, ch. 134, sec. 1; 1867, ch. 161, sec. 1; 1878, ch. 257; 1880, ch. 295; *Revised Statutes,* 1858, ch. 165, sec. 19; ch. 42, sec. 4; *Statutes,* 1898, sec. 4448–49. "Shingle bolt" was the technical term for the block from which shingles were made, as explained on page 33.

5. Milling and Marketing

INDIVIDUAL human skill counted for less in the sawmill than in the woods, but the finished product and the mill's ultimate profit could be drastically affected not only by the quality of the milling machinery but also by the competence with which it was applied to the logs. Retail lumberyards refused to buy from mills whose product was known to be of inferior quality.[1] These conditions, coupled with the growth of Wisconsin lumbering into a large-scale industry after the Civil War, resulted in the evolution of extremely efficient and speedy sawmilling techniques.

Sawmills had been operating in America at least as early as 1635, but rarely was a mill built on the frontier or in thinly settled regions. In areas where settlement had just begun it was universal practice to prepare timber for building purposes, unless the log was used in its original state, by hewing or whipsawing. Both these methods were used in Wisconsin. When a log was whipsawed it was rolled over a pit, and two men, one of whom stood beneath the log inside the pit, the other above it, cut it into boards by pulling and pushing a saw up and down. Sometimes the whipsaw was enclosed in a frame which held it taut and acted as a gauge for straight cutting.

[1] R. Musser & Co. to Ingram and Kennedy, not dated but probably written in June, 1865, in the Ingram Papers.

From this device was developed the first mechanically operated saw used in Wisconsin, the frame saw or sash saw, also occasionally called the gate saw. The early mills all used saws of this type. Its chief disadvantage was that because of its massive frame it took an undue amount of power to overcome friction. To a certain extent this difficulty was obviated before 1840 by the adoption of the muley saw, which operated exactly like the gate saw but had a lighter framework and hence could be operated at the same speed with less power.

At best, however, the frame and muley saws were extremely slow affairs. They operated with an up-and-down motion, cutting only on the down motion. With the introduction of the rotary or circular saw the productiveness of sawmills was increased nearly ten per cent with the same amount of power. This saw was invented in England sometime before 1777 and was introduced into the United States in 1814, but it was slow to find favor. Few of the early lumbermen in Wisconsin knew how to operate it, and it was not until the late fifties and early sixties that it came into general use. The rotary saw, spinning like a huge wheel, sliced into the logs that were forced edgewise against the teeth in its rim. By adding a small rotary above a larger one, logs of any size could be cut in a single operation. Development of the gang saw, which had been known in Europe as early as the sixteenth century, enabled a mill to cut dimension lumber with comparatively little waste in sawdust. The gang resembled the old-style gate saw, but it was much lighter and held a number of blades which could cut a log into boards in one operation. Its improvement kept pace with that of the circular saw. The latter was often used only to cut two slabs from the sides of a log preparatory to running it into a gang.

The rotary saw was developed to increase speed and save labor rather than to conserve raw material. Before 1880, when the legend of the inexhaustibility of American forests was still generally accepted, economy of raw material was not an important consideration. No one objected to the circular saw because it cut a half-inch kerf. It was enough that it speeded up production. During the last two decades of the century, however, it came to be realized that the forests were far from inexhaustible. Men with large timber holdings began to think of saving what they owned, and buyers of timber, faced with the prospect of advancing prices, became desirous of producing as much as possible from their logs. The result was the adoption of another refinement, the band saw. Invented by an Englishman in 1808, the band saw immediately became popular for scroll work, but in America it was not used for sawing logs until 1872. After 1880 it appealed strongly to the lumbermen of Wisconsin because it cut a thin kerf, sometimes no thicker than a knife blade, and by the 1890's all the best-equipped mills were using it. The band saw was a continuous strip of metal about fifteen inches wide and fifty feet long with teeth in one edge. It was operated like a pulley belt. Single and double band saws were more often employed to prepare logs for the gang saws than to cut them into lumber directly. After a band saw had been installed in a mill the old circular saw, instead of being abandoned, was often relegated to the sawing of coarse logs.[2]

At the same time that the mills were adopting new sawing methods, their source of power was likewise undergoing a revolution. In Europe both wind and water had been used as motive power for saws, but America's early mills were all powered by water. The first mills in Wisconsin used either an overshot or an undershot wheel. The overshot wheel was used when a comparatively great head of water could be obtained and the supply was limited, the undershot wheel when it was necessary to obtain rapid motion at the expense of a larger volume of water. Either kind of wheel was attached directly to the sash or muley saw by a heavy crank. To get a fairly good speed for the saw the wheel had to be of small diameter. When for some reason there was a backwater below the mill dam, both these types of wheel were useless, for the dead water in which the

[2] Usher, *Wisconsin*, 1:162–163; Thomas E. Randall, "Chippewa Valley History," in the *Eau Claire Free Press*, September 10, 1874; Fernow, "American Lumber," in Depew, *American Commerce*, 1:202–203; Merk, *Economic History of Wisconsin during the Civil War Decade*, 69–70; Augustus Cole to Holt and Balcom, April 19, 1887, in the Holt Lumber Company Papers. The term "muley saw" was a corruption of the German word for sawmill, *Sägemühle*.

WHi(X3)20997

Schematic drawing of a band saw and accessory equipment employed in the logging industry and manufactured by Edw. P. Allis & Co of Milwaukee, Wisconsin. Such saws increased the efficiency with which boards could be cut from logs since the width of the saw cut was much narrower than with circular saw blades.

WHi(X3)12370

Chippewa Falls, Wisconsin, probably about 1900. Interior of the "Big Mill" of the Chippewa Lumber & Boom Company, showing a large log that has already been squared off on three sides being moved through the band saw on the steam-carrier. Photo by A.A. Bish, Chippewa Falls, Wisconsin.

wheel was immersed counteracted the power. After 1850 most mills operating by water power used a turbine wheel, which was based on the principle that water forced out of a small opening in a flume causes a reaction opposite to the direction of its release. The turbine wheel could be operated in dead water if sufficient head could be furnished, and it derived the maximum power from a given quantity of water. The steam engine was introduced into sawmills before 1860, but the water turbine was efficient enough to maintain itself in Wisconsin as long as lumbering lasted.[3]

Other mill improvements followed these changes. The friction feed, wire feed, and direct steam feed increased the speed of the carriage that transported the log to the saws. One of the best of the steam feeds was developed by a Marinette inventor in 1873. Accurately adjusted setting works, head blocks and dogs, a "steam nigger" to turn the log for the saws, and endless chains to bring logs from the boom into the mill all helped to eliminate expensive manual labor. The endless-chain method of bringing logs into the mill was patented by a La Crosse lumberman in 1873. Later a device was invented for washing sand from the logs as they were being taken to the saws. Mechanical carriers for disposing of the finished lumber and of refuse were introduced, such as the invention perfected on the Chippewa River in 1869 which automatically carried sawdust to the boilers of the steam engine, where it was used as fuel. During the sixties the double-edger and gang-edger were introduced, increasing the speed with which a board could be finished. The development of shingle saws, lath saws, and slab saws made it possible to utilize the poorer materials and thus reduce waste.[4]

It was only with difficulty, however, that the practice of sawing shingles was introduced into Wisconsin because of the prejudice that existed against their production by mechanical methods. In the forties and fifties most shingles were made by

[3] Randall, "Chippewa Valley History," in the *Eau Claire Free Press*, September 10, 1874.

[4] Fernow, "American Lumber," in Depew, *American Commerce*, 1:202–203; Merk, *Economic History of Wisconsin during the Civil War Decade*, 70–72; Bruncken, *Forests and Forestry*, 79.

hand, although in La Crosse a machine was used for cutting them with a knife. In time, however, Yankee immigrants brought to Wisconsin the shingle-cutting techniques used in the East, and the old prejudice disappeared. By 1865 many mills were using the improved Challoner machine, which could saw over six thousand shingles an hour without requiring the operator to come within six feet of the saw.[5]

These improvements made their appearance in quick succession, a single year sometimes marking the introduction of several. Equipment displayed at the Wisconsin State Fair in 1873 included head blocks and setworks, a hub-capping machine, a three-knife cylinder plane, a band saw, a panel raiser, and various steam engines. The building of railroads in the West produced further changes in the industry which hastened progress. They made it possible to ship millwork directly to consuming areas and thus stimulated the introduction of planing, flooring, matching, and molding machines in Wisconsin. Many mills also began using the process of artificial drying, perfected after 1867, to prepare their lumber for sale directly to the consumer. Railroads also changed the location of sawmills. Small portable mills were used as early as 1867, but it was not until the railway made possible the direct transportation of the finished product to markets all over the West that it became more economical to build the mill as close as possible to the place of cutting.[6]

The increased efficiency made possible by the new machinery was reflected in the production of the mills. In 1860 the average Wisconsin mill produced about fifty thousand board feet of sawed lumber in a twelve-hour day, exclusive of shingles, lath, pickets, and other smaller products. Seven years later a mill in Chippewa Falls could produce from one to two hundred thousand board feet in a day. The yearly production of the ordinary

[5] Stephenson, *Recollections,* 82; Albert H. Sanford, "The Beginnings of a Great Industry at La Crosse," in the *Wisconsin Magazine of History,* 18:378 (June, 1935); Oshkosh *Northwestern,* November 9, 1865.

[6] Fernow, "American Lumber," in Depew, *American Commerce,* 1:203; Brunck-en, *Forests and Forestry,* 76–79; J. G. Ingram to Ingram, Shaw & Co., January 15, 1867, in the Ingram Papers; La Crosse *Republican and Leader,* June 2, 1877; *Wisconsin Lumberman,* 1:32–40 (October, 1873).

mill increased accordingly. In 1870 several Oshkosh mills were producing from three to five million board feet annually. Three years later the average mill could saw about thirty million feet in a year, and a few mills were producing from fifty-five to seventy million board feet. Similarly, the output of each district increased from year to year. The annual production of the Chippewa valley mounted from about forty million feet in 1858 to more than two hundred million feet in 1875 and twice that quantity in 1884. The Black River output increased from six million feet in 1858 to twenty-nine million feet in 1869 and sixty-two million feet in 1878. Oshkosh produced forty-eight million feet in 1864 and more than double that amount ten years later.[7]

The early sawmills produced only rough lumber. Before railroads were built in the pineries few lumbermen seasoned and finished their product before sending it to the markets of Chicago, northern Illinois, or points along the Mississippi, since transportation by water inevitably damaged finished lumber. This of course meant the loss of the profit that could have been made by finishing the product; but the industry was so dependent on cheap transportation that even when rail transportation became available it could never afford to abandon entirely the rafting of boards by river and lake. Faced with a choice between rail and water transportation, a company had to compare the respective costs before coming to a decision. Lumber being fairly heavy in relation to its value, the cost of transportation, whether by rail or by water, represented a large part of the market price. No lumbering concern could survive long without cheap transportation.

Indeed transportation was a more important factor in ultimate profits than any other phase of the productive process. Other things being equal, even the poorest grades of lumber could be marketed at a good profit if cheap transportation was

[7] Merk, *Economic History of Wisconsin during the Civil War Decade*, 72; *Chippewa Union and Times*, August 24, 1867; Oshkosh *Northwestern*, November 30, 1865, April 6, 1871; *Chippewa Herald*, January 9, 1874; *Northwestern Lumberman*, December 23, 1876, January 6, 1877, March 30, 1878; *La Crosse Independent Republican*, May 5, 1858; *La Crosse Leader*, January 22, 1870; *Eau Claire Free Press*, January 15, 1885.

available, whereas the return from far better grades would not cover the cost of sending them to market under unfavorable conditions. In the early years rivers and lakes provided the cheapest and most convenient means of shipment, and the pioneer lumberman quickly developed great skill in bringing his product to market cheaply by water. Before 1850 lumber rafts were floating on all the great logging streams, and Lake Michigan was carrying cargoes of lumber and logs to Chicago.[8]

All lumbermen who used the rivers for transportation constructed rafts of their lumber, although the method of construction differed somewhat on the several rivers. Usually the raft was put together in a rafting shed behind the mill and normally conformed to a standard pattern. Taking two sixteen-foot planks to form the foundation, the raft-contruction crew bored holes at the ends and middle of them. Through these holes they drove hardwood "grub pins" three feet long. They then laid boards at right angles across the foundation planks, and across these again they placed other boards until they had built up from twelve to twenty layers. Over the last layer they placed binding planks with holes to correspond with those in the foundation planks. The grub pins were then drawn up and tightened above the binding planks. This process completed a "crib" sixteen feet long and from twelve to sixteen feet wide. Six or more cribs formed a "rapids piece" or "string," which could be easily taken apart when dangerous rapids or dams had to be passed. On a large quiet stream like the Mississippi five or more strings were coupled together side by side to form a single raft.[9]

Early lumber camps shipped their hewn timber to market by a method similar to the brail system of rafting logs except that

[8] Bureau of Corporations, United States Department of Commerce, *The Lumber Industry* (3 vols., 4 parts, Washington, 1913, 1914), pt. 1, pp. 84–86; Blair, *Raft Pilot's Log*, 33–36; Hotchkiss, *Lumber and Forest Industry*, 500; *La Crosse Tribune and Leader-Press*, February 27, 1938. The best study of river rafting in Wisconsin is Wilbur H. Glover's "Lumber Rafting on the Wisconsin River," in the *Wisconsin Magazine of History*, 25:155–177, 308–324 (December, 1941; March, 1942).

[9] Bentley, Wisconsin River Lumber Rafting Days (MS.), 5–9; Lincoln, Lumbering in Early Days (MS.), 6–8; Ellis, "Upper Wisconsin Country," in *Wisconsin Historical Collections*, 3:441–443; Merk, *Economic History of Wisconsin during the Civil War Decade*, 81–83.

binding poles were laid across the frame and fastened to the timbers by thin strips of tough oak sapling. A method something like this was used around Oshkosh for lumber rafts. When the railroads had advanced far enough into the West to furnish a good market for wood products, ties were carried to Mississippi points in the form of rafts constructed by nailing the ties across two parallel poles.[10]

The size of the lumber rafts increased with the industry. Before 1860 the larger of the Mississippi River rafts comprised from forty to fifty cribs, containing from three to five hundred thousand board feet of lumber. By 1864 some rafts were carrying as much as a million feet, and by 1870 more than double that amount. In that year a Chippewa Falls mill sent to St. Louis a raft of two and a half million feet, enough to build a village of a hundred and twenty-five houses. Such rafts had an area of three or four acres. During the decade of the seventies rafts composed of three million feet were reported, although the average size was much less—from three-quarters of a million to a million and a half board feet. In 1889 a raft of three and a half million feet was still a novelty, but in 1895 seven million feet of lumber was carried down the Mississippi in a single raft. It was the usual practice to top the lumber rafts with deckloads of shingles, lath, and pickets. Occasionally a small operator shipped his lath and shingles on the rafts of larger mills.[11]

Until the mid-sixties all rafts were floated downriver by the current and kept in the channel by crews who managed them at bow and stern with sweeps about thirty feet long. Pulling a raft oar required skill and great strength. The operator dipped

[10] Vinette, Reminiscences (MS.), 6; *Northwestern Lumberman,* April 5, 1879; Harry Dyer, Tie Rafting, manuscript dated 1931, and David G. James, Tie Rafting, manuscript dated 1912, in the Brown Papers.

[11] Knapp Dairy, 1859; *Chippewa Falls Democrat,* August 12, 1869, June 9, 1870, March 3, 1871; *Chippewa Herald,* November 9, 1872, June 7, 1895; La Crosse *Republican and Leader,* December 29, 1877, July 26, 1879; *Eau Claire Free Press,* July 31, 1879, June 11, 1889, June 6, 1895; *Northwestern Lumberman,* August 2, 1879; W. B. Esterbrooks to Orrin H. Ingram, October 1, 1858, in the Ingram Papers. A federal geologist reported in 1848 that the steamer *War Eagle* towed out of Lake St. Croix a raft of logs and lumber which covered eleven acres "by measurement." Owen, *Geological Reconnoissance of the Chippewa Land District,* 71.

the oar by swinging it at arm's length above his head and then, walking in a short arc, surging upon it at each step until the stroke was completed. The men who performed this work considered themselves the aristocracy of labor in the industry. Contemptuous of danger, carefree, footloose, and picturesque, they looked down upon men whose profession they considered less worthy. They could always be distinguished by their lack of restraint in dress and manner. River towns dreaded their approach, but with their songs and dances they must have injected a touch of romance into the lives of those who saw them. During the trip each man built on the raft a small shanty for his sleeping quarters, and a cook shanty was built in the center. Each of the rafts was in charge of a pilot, who must know every inch of the river. On the Mississippi they sometimes ran day and night, but it was customary to tie up to the bank at sundown.

To stop a huge lumber raft was no easy task. When it was decided to tie up for the night, one of the crew jumped ashore with a cable, which he made fast with a peculiar hitch to a tree or post, while several men on the raft played out the rope slowly so that the raft's momentum would not snap the line. To make the playing out, or "rendering," easier, a snubbing device was built near the end of the raft. This consisted of one or two logs fastened at right angles across another big stick which was made fast to the deck. By twisting the cable around the ends of the logs, enough friction could be created to permit the line to be let out slowly, but a man had to keep pouring water over the rope to put out the fire generated by the friction. If the cable broke, the crew resorted to "growsing" by shoving planks or poles down between the boards of the deck so that they scraped the bottom of the river. At St. Louis a snubbing works had been set up on shore to tie up the many rafts that ended their journey there. When a raft destined for that city reached Alton, Illinois, it waited to receive advice from St. Louis when it could be tied up at the snubbing works.[12]

To get the crews back home after the raft had reached its

[12] Lincoln, Lumbering in Early Days (MS.), 7, 17, 18, 21–22; Note on Snubbing Device, in the Brown Papers; Blair, *Raft Pilot's Log*, 34.

destination was always a problem. Early crews, who took rafts only short distances, sometimes returned on foot. Wagon roads and stage lines were used whenever possible, and experiments were made with steamboat transportation. After the steam towboat came into common use, many crews returned upriver with the steamer. By the 1870's railroads began to be used for the purpose.[13]

The use of steamers for towing rafts worked great changes in the industry, cutting in half as it did the time required to take a raft to market. Towing of rafts had begun as early as 1848, but it was not until after 1864 that its efficiency was fully demonstrated. By 1873 seventy-three steamboats were engaged in towing on the upper Mississippi, and at the peak of the industry in Wisconsin more than a hundred were so engaged. Some lumber companies had their own towboats, but usually the boats were hired by the trip. The steamer was always placed at the stern of the raft to facilitate the management of the unwieldy craft. Steering was managed by changing the angle of the push and by using sweeps at the bow. As ever-larger rafts were assembled it became common practice to place a small steamer at the bow of the raft at right angles to the current. This boat could swing the raft in any direction. When a raft was towed the raft pilot's work was taken over by the steamboat pilot. In 1888 an upper Mississippi River Pilots' Association was organized for joint discussion of the difficulties of navigation and cooperative handling of common problems. The federal government set up agencies for licensing and inspecting towboats at important points along the river.[14]

The entry of the towboat into the lumber-rafting field brought to an end the picturesque aspects of the rafting crews, largely because it deprived them of the time they had spent in

[13] *Assembly Journal,* 1871, Appendix, 38–39; *Chippewa Herald,* August 10, 1872; *Eau Claire Free Press,* November 15, 1883.

[14] Owen, *Geological Reconnoissance of the Chippewa Land District,* 71; Knapp Diary, 1859; Holley, "Waterways and Lumber Interests," in *Proceedings of the State Historical Society of Wisconsin,* 1906, p. 214; Merk, *Economic History of Wisconsin during the Civil War Decade,* 84; *Chippewa Falls Democrat,* September 2, 1869; La Crosse *Chronicle,* January 17, December 5, 1889; Logs and Lumber Rafting from Beef Slough, 1872–1875, manuscript in the Brown Papers.

play as their raft dawdled along with the current. In 1874 a lumber journal was complaining that "the romance has gone out of lumbering to a great degree. The raftsmen who were handed down to posterity as living in the utmost contentment on the raft, and passing life in an unbroken round of cardplaying, fiddling, and dancing while they floated lazily down the stream, have taken their places at the titanic oar and a noisy raft-boat, powerful enough to haul an island out of the way, pushes the monstrous raft to its destination."[15]

Lumber districts tributary to Lake Michigan did not use the same methods of transportation as the Mississippi River area, though both logs and lumber were rafted successfully on the lake. The Green Bay region depended on lake boats to bring in its supplies and carry out its lumber. Before 1860 the old sailing type of vessel was used for this purpose; frequently thirty or more of these small boats were anchored outside the bar at the mouth of the Menominee River. But dependence on wind power made marketing an uncertain business at best. After a long spell of prevailing southerly winds the port of Chicago was likely to be empty of lumber cargoes until the wind changed, when the Chicago River would again become choked with lumber ships all the way from the Lake Street to the Clark Street bridge.[16]

The arrival of the railroad at the lake ports changed the character of navigation on the lakes. It captured so much of the carrying trade that many good passenger and cargo vessels were forced out of business. When in 1861 a Buffalo lumber merchant discovered that they could be dismantled and towed as barges by steam tugs, he hit upon a scheme that was adopted all over the lake area, despite the fact that the ordinary barge of this type carried hardly more than three hundred thousand board feet of lumber and a quantity of smaller products like lath. In 1870 a new epoch in lake transportation was inaugu-

[15] Gregory, *West Central Wisconsin,* 1:219.

[16] La Crosse *Republican and Leader,* July 5, 1873; Stephenson, *Recollections,* 92; Hotchkiss, *Lumber and Forest Industry,* 297; Merk, *Economic History of Wisconsin during the Civil War Decade,* 79–80; *American Lumberman,* December 17, 1938 (clipping in the Holt Lumber Company Papers).

rated when propellers capable of carrying cargo as well as supplying motive power began to replace the former tugs. The Marinette Barge Line Transportation Company, formed in 1873 with a propeller and twelve barges having a capacity of three million board feet apiece, was the most notable example of this type of transportation in Wisconsin. By 1897 most of the old lake schooners had been transformed into barges for the lumber traffic. As a matter of fact, the lumber fleet on the lakes was largely made up of vessels useless for other purposes. Boats that had outlived their serviceability as ore or flour carriers could still be employed by the lumber industry for years, for lumber kept a water-logged vessel afloat and was less damaged by contact with water than a flour cargo, for example.[17]

Because of the federal government's tardiness in improving the lake harbors it was for many years needlessly difficult to load a lake vessel with a lumber cargo, for the cargo had to be towed out to deep water in small rafts or barges. But ultimately the government did lend some assistance. In response to the pressure of the lumber companies around Peshtigo it made a land grant to a private company for the construction of the Sturgeon Bay Canal, which upon its completion in 1873 eliminated the slow and dangerous detour through the Death's Door passage; and twenty years later it bought the improvement for about a hundred thousand dollars.

Both natural and man-made obstacles made the rafting of lumber hazardous to property and life. An experienced raftsman listed as the most serious perils on the Wisconsin River, a stream noted for the numerous risks it presented, the following: Big and Little Bull Falls, the Stevens Point dam, Courant's Rapids, Grand Rapids, Clint's Big and Little Dam, Whitney's Rapids, the Dells, Kilbourn Dam, "and a great many other places." The Little Bull Rapids at Mosinee were considered the worst spot

[17] George G. Tunell, *Statistics of Lake Commerce* (55 Congress, 2 session, House Document no. 277, serial 3679, Washington, 1898); Hotchkiss, *Lumber and Forest Industry,* 297; Accounts of Lumber Shipped during the Summers of 1867–1871 by Holt and Balcom of Oconto, Wisconsin, in the Holt Lumber Company Papers; *Waukegan Harbor and Oconto and Wolf Rivers* (46 Congress, 2 session, House Executive Document no. 19, serial 1913, Washington, 1881); Stephenson, *Recollections,* 92, 168, 171–173.

on the river. There the stream narrowed to a width of only thirty feet, and for half a mile the water plunged in a swirling flood through a gulch of rocks thirty feet high. For the protection of the crew a "sucker" line to which the men could cling was stretched across the length of the raft before it reached the narrows. Frequently a raft drove so deeply into the water that only the heads of the crew remained visible, and sometimes they were completely submerged.

The other rafting streams also had their dangers. The greatest hazard on the Chippewa was the "falls," where the water rippled through a long series of rapids and eddies. It was because the falls were so risky that most of the lumber milling on the river became concentrated below them. On the Black River the raftsmen worried over Snag Bend and Gibbs Chute, but their greatest problem was low water. The Mississippi was comparatively safe for lumber rafts, but a sudden recession of the water sometimes left one of them stranded on dry land thirty rods or further from the bank. After the high water of June had receded, one could see along the bank tree trunks from which the bark had been stripped almost to the top by the bobbing rafts.

Dams and other man-made obstructions endangered the lumber rafts quite as much as did falls and rapids. Every dam of any size had an inclined chute over which rafts could slide to the water level below, but no amount of engineering skill could provide a raft chute that was absolutely safe. In going over dams much lumber was shaken from the raft and lost. There were those who made a profession of waiting below such dangerous places to gather up the boards and shingles lost by unfortunate raftsmen.[18]

To protect the owners of lumber rafts the state law provided that navigable streams must be kept open to traffic at all times and that anyone placing obstructions in them was liable for damages. Federal law and precedent had positive injunctions of a similar nature. The charters of dam companies, too, usually

[18] Lincoln, Lumbering in Early Days (MS.), 8, 13–14, 20; Bentley, Wisconsin River Lumber Rafting Days (MS.), 1; Vinette, Reminiscences (MS.), 6; *La Crosse Tribune and Leader-Press,* September 20, October 4, 1936; *Assembly Journal,* 1871, Appendix, 42 and *passim.*

contained clauses making them liable for injuries to lumber rafts caused by their improvements. Lumbermen frequently took advantage of such clauses to sue for the recovery of losses from rafts, but litigation was a long, unsatisfactory process to men who were almost always short of liquid capital. Occasionally several firms sued jointly for the removal of obstructions charged to be nuisances. But in the long run the best protection was a sturdy and strong raft and the placement of heavy timbers where they would absorb most of the shocks it had to sustain.[19]

The rafting interests naturally looked to the state and federal governments to keep the rivers in good condition. In 1852 the legislature petitioned Congress for a grant of the odd-numbered sections of land on each side of the Wisconsin River from Winnebago Portage to Beaulieux Rapids to be held in trust until "the State shall sell such lands and apply the nett proceeds to the improvement of the upper Wisconsin River." From 1853 to the end of the century the state gave the federal government the right to acquire property for purposes of river improvement. Army engineers did much to improve rafting conditions throughout the state by removing natural hazards, but, as they reported to Washington, so long as the lumbermen themselves continued to build "improvements" that were detrimental to rafting, it was almost useless to spend money removing other dangers.[20]

Railroad bridges were a particularly annoying hazard to the rafting interests. Rarely did the railroads when they built their bridges take account of the needs of the rafting industry, and almost every year many rafts disintegrated while trying to squeeze between bridge piers set too close together to allow safe

[19] *Revised Statutes,* 1858, ch. 19, sec. 112; *Private and Local Laws,* 1861, ch. 285, sec. 1; 1866, ch. 123; 1869, ch. 152; *Statutes,* 1898, sec. 1598, 1599; *Supplement to the Wisconsin Statutes of 1898,* sec. 1598; B. E. Reid to Orrin H. Ingram, May 12, 1875, and Knapp, Stout & Co. to Ingram, Kennedy & Co., September 10, 1872, in the Ingram Papers; Usher, *Wisconsin,* 1:171.

[20] *Memorial of the Legislature of Wisconsin Praying a Grant of Land to Aid in the Improvement of the Wisconsin River between Fort Winnebago and the Beaulieux Rapids* (32 Congress, 1 session, Senate Miscellaneous Document no. 73, serial 629, Washington, 1852); *General Laws,* 1853, ch. 72, sec. 2; *Revised Statutes,* 1858, ch. 41, sec. 2; *Session Laws of Wisconsin,* 1895, ch. 372; *Statutes,* 1898, sec. 1596; *Chippewa River, Wisconsin* (42 Congress, 2 session, House Executive Document no. 278, serial 1516, Washington, 1873); La Crosse *Chronicle,* January 22, 1880; *Eau Claire Free Press,* October 7, 1886.

passage. In 1876 six rafts broke up in a single day at the Milwaukee and St. Paul railway bridge near La Crosse, and "irrepressible war" between the railway and the rivermen was launched. In desperation rafting interests and steamboat companies urged the federal government to declare dangerous bridges "obstructions to navigation." Eventually, despite the influence of the railroad corporations in Congress after 1870, the rivermen won their fight, probably because law and precedent were on the side of navigation. After 1880 statutes authorizing the construction of bridges across navigable rivers required the builders to provide adequate safeguards for rafts and other river craft, and army engineers consulted lumbermen to learn whether any bridges were "regarded by your company, or by any other lumber companies . . . as an obstruction to rafts and steamers."[21]

Lumber rafts on the Great Lakes caused the government much trouble during the eighties and nineties. Huge rafts, six or seven hundred feet wide, passing through the narrow channel between Lake Superior and Lake Huron caused collisions, displaced marking buoys, and in general disrupted and slowed up navigation. As a result the government laid down strict regulations governing the size of lake rafts, their proper lighting, the tugs' handling of them, and the restoration of displaced buoys.[22]

River rafting was a hard and dangerous life, but it had "a fascination that held men in spite of the danger to life and limb." After every hard season many raftsmen "quit the river," but the "first breath of Spring was Sure to bring the fever." The type of adventurer who was attracted to such a life could not be discouraged by mere peril.[23]

The raft pilot, because of the responsibility he shouldered, commanded in ordinary times, during the years 1850–1900, a daily wage of four to fifteen dollars, depending upon the length and difficulty of the trip. Many pilots, however, preferred to do

[21] *Chippewa Herald,* August 18, 25, 1876; La Crosse *Republican and Leader,* March 22, 1873; Charles J. Allen to Orrin H. Ingram, May 2, 1885, in the Ingram Papers.

[22] *Lumber Rafts on Great Lakes* (51 Congress, 2 session, House Report no. 3759, serial 2888, Washington, 1891).

[23] Bentley, Wisconsin River Lumber Rafting Days (MS.), 14.

the work on contract. A rafting contract usually stipulated that the jobber construct the raft, hire the crew, pay all the expenses, and deliver the lumber at a specified market, and before 1870 the contractor was often entrusted also with the actual sale of the raft. Lumber mills that preferred to let their rafting to contractors advertised for bids and sometimes signed contracts for two years or longer. But never was all rafting done on a contract basis. So important was the safe arrival of a raft at its destination that many companies were unwilling to entrust the responsibility to any but their most faithful employees. Even when for some reason a mill did not have enough lumber on hand to build a fair-sized raft of its own, it often preferred to construct one jointly with another company rather than rely upon an independent rafting contractor.[24]

Despite the fact that before 1860 it was always cheaper to ship by water than by land, the cost of sending a raft from a Wisconsin point to a market on the Mississippi sometimes consumed nearly fifty per cent of the receipts from its sale. Three rafts sent to Mississippi River markets in 1848 by a Menomonie company brought a gross return of $5,087.34, of which $2,423.48 had to be charged to rafting expenses. In 1852 Moses M. Strong paid $619.65 in wages to have a fleet of rafts totaling 265,383 board feet conducted from Stevens Point to Arena, an average cost of almost $2.35 per thousand board feet. In 1857, when lumber was selling for about twenty dollars, the cost of running rafts from the upper Wisconsin River to Mississippi River points ranged from five to eight dollars a thousand board feet. In 1858, after the panic of 1857 had lowered prices and lumber was selling for three to five dollars, rafting expenses from Eau Claire to St. Louis amounted to $1.65 or more per thousand feet. But when lumber prices were restored to a normal level of about twelve dollars in 1860 and 1861, the cost of rafting from Eau Claire to St. Louis remained at $1.50. The Civil War drove all costs and

[24] Ellis, "Upper Wisconsin Country," in *Wisconsin Historical Collections,* 3:439; *Chippewa River, Wisconsin;* Receipt of William B. Newcomb for Raft, September 16, 1858, Receipt of Thomas Fitzgerald, October 9, 1858, and letter of De Witt C. Clark to Ingram and Kennedy, June 11, 1865, in the Ingram Papers; *Chippewa Herald,* September 7, 1872, January 11, 1873.

prices to unprecedented levels, but by 1866 normal conditions had returned.[25]

Rafting costs were directly affected by the water level and by obstructions to safe navigation. In 1872 army engineers estimated that the cost of moving a raft from Chippewa Falls to Eau Claire in high water without obstructions was seven and three-tenths cents per thousand board feet, eleven cents in low water. The presence of obstructions increased these costs to fourteen and six-tenths cents and twenty-three and a quarter cents respectively.[26]

Lumber companies and rafting contractors had other expenses besides the wages of their crews. Cities, villages, and landowners along the route of the rafts charged a fee for the privilege of tying up to the bank. In 1868 a Mississippi River village demanded $29.41 for "wharfage" on one hundred and eighty-one thousand board feet, or about 16.2 cents per thousand. River-improvement companies had the right to collect tolls from rafts passing their works. In 1874 the Wisconsin River Improvement Company set toll rates ranging from seven and a half to ten cents per thousand board feet for passing Big Bull Falls, Little Bull Falls, Grand Rapids, and Lower Rapids, a separate toll being collected at each place; shingles were assessed at two cents a thousand. In 1878 the cost of hiring a steam towboat for a thirteen-day trip came to $1,027.09. In general the rental was fixed at a figure that would bring the towboat owner a minimum net profit of six per cent on the trip.[27]

[25] *Dunn County News,* November 16, 1911; Chetek *Alert,* September 22, 1916; Thomas J. Moorman to Moses M. Strong, July 4, 1852, and Statement of Expenses, August, 1852, in the Strong Lumber Papers; Ellis, "Upper Wisconsin Country," in *Wisconsin Historical Collections,* 3:440, 443; John Welch to Dole, Ingram and Kennedy, November 11, 1858, A. M. Dole to Ingram and Kennedy, February 17, 1860, and George Winans to Orrin H. Ingram, April 14, 1861, in the Ingram Papers; *La Crosse Independent Republican,* May 19, 1858; Knapp Diary, 1860.

[26] *Chippewa River, Wisconsin.* See also *Wisconsin Assembly Journal,* 1871, Appendix, 37, which gives raft pilots' testimony that it cost twelve cents per thousand board feet to bring a raft from Chippewa Falls to Eau Claire and forty-one and a half cents from Menomonie to the Mississippi.

[27] Peter Klein to Moses M. Strong, June 15, 1868, in the Strong Lumber Papers; *Wisconsin Lumberman,* 2:15 (April, 1874); Blair, *Raft Pilot's Log,* 75.

The natural waterways of the Great Lakes system made Chicago the earliest focal point of the lumber trade in the Middle West. The city not only had easy access to the entire Green Bay and western Michigan logging regions, but was itself easily accessible to the consuming areas of both East and West. Before 1833, when it was a frontier settlement of some thirty people, its lumber was hauled by ox team from the Ohio River, but by 1835 lumber from the west side of Lake Michigan was beginning to reach the village. In 1839, when its population had grown to more than four thousand, Wisconsin lumber was arriving in considerable quantity, and by the mid-fifties it was competing on equal terms with the Michigan product.[28]

The rapid growth of Chicago and its hinterland soon made the lake port the greatest lumber mart in the world. By 1857 it was supporting more than a hundred and fifty lumber firms, having an aggregate investment of more than sixty million dollars and a payroll of ten thousand employees. In 1872 three-fourths of the lake vessels arriving at the port were loaded with lumber. After 1870 the Illinois and Michigan Canal and the railroads enabled Chicago to ship lumber supplies from Michigan, Wisconsin, and Minnesota to all parts of the country. Lumbermen attributed to their trade the main credit for the growth of Chicago into a great commercial city.[29]

Chicago was the chief market for the lumber in the Green Bay and Wolf River areas, but eastern Wisconsin sent its product to other wholesale centers also. Up to 1850 Oswego on Lake Ontario was the eastern distributing point for much of the lumber from the Green Bay region, but as Atlantic coast dealers began feeling the pinch of declining local supplies and turned to the West

[28] Dopp, "Geographical Influences in the Development of Wisconsin," in the *Bulletin of the American Geographical Society,* 45:738 (October, 1913); Hotchkiss, *Lumber and Forest Industry,* 29, 219.

[29] *Wisconsin Lumberman,* 1:12 (October, 1873); *Chippewa Herald,* January 8, 1875; *Eau Claire Free Press,* September 30, 1875; Merk, *Economic History of Wisconsin during the Civil War Decade,* 81. Lumber receipts at Chicago reveal the growth of both the lumber industry and the Middle West: 1847, 32 million board feet; 1855, 306 million; 1865, 647 million; 1875, over a billion; 1885, 1.75 billion; 1892, over 2 billion. Fernow, "American Lumber," in Depew, *American Commerce,* 1:198.

for new sources, Buffalo superseded Oswego as a lumber center. Tonawanda on the Niagara River and Toledo on Lake Erie were also important markets for eastern Wisconsin. The Wolf River district sent some of its lumber to Mississippi River markets by way of the Fox-Wisconsin route. Both Oshkosh and Fond du Lac sent lumber across Lake Horicon and down the Rock River as far as Rockford.[30]

Lumber mills in western Wisconsin sent their lumber product to points along the Mississippi River, whence it could be conveniently transported by land to the treeless prairie regions of the West. A whole series of distributing centers grew up along the river, St. Louis soon outranking all others.[31] As early as 1844 half the sawed lumber from the upper river went to St. Louis. As settlement increased in the trans-Mississippi region, Iowa river towns received and distributed more than half the lumber product of the Wisconsin and Minnesota pineries tributary to the Mississippi, a total of some four hundred and fifty million board feet in 1869. In 1873 central and northern Wisconsin shipped ninety per cent of its lumber product by way of the tributaries of the Mississippi. Virtually all this lumber was consumed west of the river. Railroad development not only opened up the trade of the trans-Mississippi region for the river towns, but it also made secure the position of St. Louis as a lumber metropolis. During the seventies it was the ambition of La Crosse to become a great primary depot into which lumber from all over the state would be drawn by a network of railroads, but St. Louis, which was in a more favorable position because of its central location, easily maintained its lead.[32]

[30] Stephenson, *Recollections,* 82; Hotchkiss, *Lumber and Forest Industry,* 350 ff.; Schafer, *Winnebago-Horicon Basin,* 269, 273, 289–290.

[31] Before the development of the Western railroad network the principal distributing points south of St. Paul were Red Wing, Wabasha, Winona, La Crosse, McGregor, Guttenberg, Dubuque, Bellevue, Savanna, Galena, Fulton, Lyons, Clinton, Moline, Rock Island, Davenport, Muscatine, Burlington, Fort Madison, Keithsburg, Oquawka, Montrose, Keokuk, Canton, Quincy, Hannibal, Louisiana, Alton, and St. Louis. Many of these became milling centers as well. Durant, "Lumbering and Steamboating," in *Minnesota Historical Collections,* vol. 10, pt. 2, p. 665; A. M. Dole to Ingram and Kennedy, December 21, 1857, in the Ingram Papers.

[32] *Trade and Commerce on the Upper Mississippi River;* E. H. Playter to

From the distributing centers on the Mississippi, Wisconsin lumber found its way into every state and territory in the Great Plains. Before 1880 an observer noted that large retail yards were selling Wisconsin lumber in Missouri, Iowa, Illinois, and Kansas, and it was also being sold in Nebraska, Texas, and Colorado. By 1900 the iron-mining and wheat-raising regions of northern Minnesota and the Dakotas were consuming large amounts of Wisconsin lumber. Occasionally it found its way even to European and West Indian markets. As the West continued to grow during the eighties and nineties and as new uses for wood were discovered, the demand for Wisconsin lumber increased at the rate of twenty-five per cent a year.[33]

Before 1870 virtually all sawings from the Wisconsin mills reached the retailer through the hands of independent wholesalers. Rafts accompanied by a salesman stopped at each distributing center, where the local wholesalers bought what they wanted. If part of the raft remained unsold when it reached St. Louis, it was turned over to an agent, who undertook to dispose of it while the salesman returned to headquarters to join another raft.[34]

Eventually the demand became great enough to permit the sale of lumber at the mill itself. By 1861 wholesale yards on the Mississippi were sending orders to the mills before the rivers were open for shipment, and soon they were sending buyers into Wisconsin to contract for lumber several months in advance of

Ingram and Kennedy, September 15, 1868, in the Ingram Papers; *Wisconsin Lumberman,* 1:2 (October, 1873); Merk, *Economic History of Wisconsin during the Civil War Decade,* 85; *La Crosse Leader,* July 23, 1870; *Eau Claire Free Press,* January 22, 1874.

[33] On the markets of Wisconsin lumbermen see, for example, John A. Warder, "Forests and Forestry in Wisconsin," in *Transactions of the Wisconsin State Horticultural Society,* 1880–81, p. 153; *Wisconsin Lumberman,* 2:358 (July, 1874); Oshkosh *Labor Advocate,* September 22, 1893; *Eau Claire Free Press,* February 7, 1887; *Northwestern Lumberman,* February 17, September 15, October 20, 1877, January 26, 1878. For data on uses of Wisconsin lumber and increasing demand, see Hotchkiss, *Lumber and Forest Industry,* 752; Ellis "Upper Wisconsin Country," in *Wisconsin Historical Collections,* 3:439; Merk, *Economic History of Wisconsin during the Civil War Decade,* 60; *Wisconsin Lumberman,* 1:23 (October, 1873); 2:142 (May, 1874); Bureau of Corporations, *The Lumber Industry,* pt. 1, p. 36.

[34] Knapp Diary, 1848–1861; E. H. Playter to Orrin H. Ingram, July 3, 1871, and G. A. Porter to Ingram, Kennedy & Co., April 27, 1875, in the Ingram Papers.

its sawing. The competition between mills always remained strong enough, however, to necessitate the employment of salesmen. Besides, the sales personnel performed functions other than selling. Where lumber was shipped by lake barge it was customary for a salesman to send loading instructions before the barge took on cargo.[35]

The lumber wholesaler occupied a peculiar position in the economy of the industry. Unlike the millowner, who was chronically short of liquid assets, especially in the early years, the middleman could manage to accumulate some unencumbered cash. Not infrequently he loaned money to needy mill men, taking as his security the lumber product of the next season. Having thus placed the millowner in a defensive position, he could buy his lumber cheap and sell it at a good price, whereas the mill company was deprived of the opportunity to hold its goods for the most favorable price. Even millowners who were not indebted to wholesalers were in an unfavorable position if all the wholesalers in a market area chose to offer low prices. These circumstances led many millowners to set up their own wholesale and retail yards at the chief distributing points.[36]

Another factor contributed to this development. Some lumbermen believed that the best profits lay in the retail trade. As one of them wrote to his partners in 1860: "The retail dealers in my opinion are the ones who has made their money cheapest and easiest as well as has made the best investments. . . . I must still think Pileing and dressing our best lumber before its sale is the most profitable course." With this incentive, most of the

[35] D. T. Robinson to Ingram and Kennedy, December 29, 1871, and R. Musser & Co. to Dole, Ingram and Kennedy, January 7, 1861, in the Ingram Papers; Augustus Cole to Holt and Balcom, July 12, 1887, in the Holt Lumber Company Papers; *Chippewa Herald,* April 29, 1881; *Eau Claire Free Press,* February 15, 1883. The relationship between wholesaler and millowner was usually an intimate one. Advance orders were often contained in long, gossipy letters that detailed useful information on the state of the market, enabling the millowner to decide upon the amount of his sawing with some regard for the probable demand.

[36] M. C. Slutts to Moses M. Strong, August 22, 1865, in the Strong Lumber Papers; Merk, *Economic History of Wisconsin during the Civil War Decade,* 88–89; *American Lumberman,* December 17, 1938 (clipping in the Holt Lumber Company Papers).

large mills established their own yards before 1860. St. Louis, the terminus on the Mississippi where unsold rafts had to be held until a buyer was found, was the first choice of most western Wisconsin companies who set up their own yards. An Oshkosh firm had a yard at Rockford in 1865, and mills in the Green Bay region maintained yards in Chicago and Milwaukee. By 1874 most of the mills shipping down the Mississippi had yards in the larger distributing centers.[37]

To simplify the problem of managing their downriver yards, some milling firms formed a partnership in each yard with a local businessman. This system led to another step in the direction of expansion. Since it was often more profitable to send logs downriver than to send lumber—the price of logs being relatively higher—some of the yard partnerships constructed sawmills at the sites of their marketing plants. Other mills were built at downriver points by independent firms. Besides maintaining mills close to the market, many companies set up yards in connection with their home mills. Some Wisconsin cities, regarding the geographical expansion of lumbering as a drain on local capital, complained of the rapid extension of wholesale and retail yards.[38]

The best customers of retail yards were the farmers. When the price of wheat was abnormally low or when a grasshopper plague struck the farms of the Middle West, the lumber industry suffered. Even so, lumbermen and farmers could often strike a favorable bargain on a barter basis when money was scarce, for

[37] A. M. Dole to Ingram and Kennedy, December 17, 1860, and James Roberts to Dole, Ingram and Kennedy, January 30, 1861, in the Ingram Papers; Knapp Diary, 1860; Oshkosh *Northwestern,* August 17, 1865; Stephenson, *Recollections,* 82–83; *Wisconsin Lumberman,* 2:168 (May, 1874); *Chippewa Herald,* October 23, 1874.

[38] W. H. Day to Orrin H. Ingram, July 19, 1869, W. V. Gill to Ingram and Kennedy, July 28, 1869, and W. H. Day to Ingram, January 14, 1869, in the Ingram Papers; *Eau Claire Free Press,* December 23, 1875, March 15, 1883. Dubuque, Clinton, Davenport, Burlington, Rock Island, Keokuk, Quincy, Hannibal, Alton, and St. Louis became milling centers. La Crosse *Republican and Leader,* February 8, 1873. It is strange that La Crosse should have complained in 1854 of the lack of a yard at which it could buy lumber when a steam mill was operating every day. *La Crosse Independent Republican,* August 23, 1854.

the mill companies, with their stores and logging crews, were heavy consumers of farm products.[39]

Competition among lumber companies for the trade of "country customers" was keen, and the cutthroat price wars that developed as a result forced the industry eventually to adopt price-fixing agreements. In 1861 the agent of an Eau Claire firm that was trying to sell its lumber in a St. Louis yard for fifteen dollars per thousand board feet reported that one of its competitors had offered farmers a price of thirteen dollars, and another would give three months' credit for a price of fourteen dollars. "This will give you an Idea of how difficult it is to sell to a country customer," he wrote plaintively, "and they all have to pass both yards to come to yours."[40]

During the 1880's the yellow pine of the South began to compete with Wisconsin white pine for a place in the market. The South had an advantage over Wisconsin in that both stumpage and labor were cheaper there and the Southern railways were willing to arrange convenient rates for shipping lumber northward. The result was that Southern pine sometimes sold in Wisconsin for less than the home product. Wisconsin lumbermen met the challenge by organizing their industry as well as they could to cut down competition at home, and then, to play doubly safe, they hedged by investing in Southern timberlands.[41]

Regional competition also existed. By 1861 the wholesale lumber dealers of Chicago had railway connections with the Mississippi River, and river dealers who had been getting their supply from Wisconsin mills were complaining that Chicago could undersell them. Many locally owned yards along the river, protesting that they could not compete with yards owned by Wisconsin mills, turned spitefully to Chicago for supplies. Dur-

[39] Henry F. Black to Ingram and Kennedy, February 10, 1869, in the Ingram Papers; *Eau Claire Free Press,* March 8, 1883; La Crosse *Republican and Leader,* May 8, 1875.

[40] James Roberts to Dole, Ingram and Kennedy, February 18, 1861, in the Ingram Papers. On price-setting see also letters to Ingram and Kennedy from R. Musser & Co., June 12, 1865, J. B. Randall, April 17, 1865, and W. V. Gill, April 22, 1865, likewise in the Ingram Papers.

[41] *Eau Claire Free Press,* February 25, 1886, August 27, 1889; Cameron, *Governmental Forest Control,* 193.

ing the Civil War especially, when lumber prices on the Mississippi soared to unprecedented heights, lumber centers on the river could buy more cheaply from Chicago than from Wisconsin mills. Chicago dealers endeavored to build up a trade along the Mississippi by cutting their prices, and the railroads helped them by offering reduced freight rates to the river and beyond.[42]

[42] James Roberts to Dole, Ingram and Kennedy, January 30, 1861, R. Musser & Co. to Ingram and Kennedy, June 12, 1865, and June 27, 1865, and D. R. Moon to Orrin H. Ingram, October 28, 1865, in the Ingram Papers.

6. Railroads and the Lumber Industry

THE swift development of Western railroads in the sixties and seventies revolutionized the lumber industry of Wisconsin. So long as lumbermen remained dependent on water transportation their business was limited both in volume of production and in the speed with which raw material and finished article could be moved. Camps, mills, and wholesale yards must be conveniently accessible to navigable streams, and any thought of expansion into many inviting fields must be dismissed as impractical. The railroads enabled the industry to exploit new timber resources and develop new markets. They speeded up the processes of production and marketing, and released lumbermen from their dependence upon so uncertain an element as the weather.

The realization of these advantages depended upon two things: the construction of lines connecting the producing and consuming areas, and freight rates comparing favorably with the cost of water transportation. Before 1865 the lumbermen of Wisconsin were too content with rafting to give much consideration to the advantages that the railroads could offer them. As a matter of fact, their satisfaction with river transportation accounts in some measure for Wisconsin's failure to keep pace with Western railroad building in general during the decade of the sixties.[1]

The rapidly expanding market after the war, however, convinced most lumbermen that dependence on river transportation was retarding the industry. After 1870 desirable timber close to driving streams became ever scarcer, and loggers began speculating on means of exploiting large tracts that were practically valueless without means of transportation to the mill. At the same time certain millowners were discovering the economy of building their mills close to their timber supply and thus overcoming the uncertainties inherent in water transportation of their logs.[2]

Economical marketing, too, called for better transportation facilities. Certain kinds of timber when floated in the turbid waters of the chief rafting streams reached the market watersoaked and covered with mud. Moreover, the river town to which the lumber had been consigned was rarely its ultimate destination. By about 1870 many lumbermen had become convinced that rail transportation direct to the consumer would enhance profits. The cost of shipping lumber from Wisconsin straight to western Iowa, Nebraska, or Kansas, they told themselves, should certainly not exceed the cost of sending it to Chicago or St. Louis and from there reshipping it to the West. After all, Omaha was only sixty-one miles farther from Stevens Point than from St. Louis. The added profit to be derived from selling lumber in good condition could not be ignored, and very attractive was the prospect of eliminating the middleman by transferring wholesale centers to the mills. Furthermore, expansion into such fields as the manufacture of finished lumber, sashes and doors, tubs, furniture, and other wooden articles was largely contingent upon the development of transportation to markets. The lumbermen of the Chippewa valley especially, from whom the Mississippi milling towns received most of their logs, were

[1] Orrin H. Ingram, *Letters of a Pioneer* (Eau Claire, 1916), 6; A. M. Dole to Ingram and Kennedy, January 9, 1858, in the Ingram Papers; Carl Russell Fish, "Phases of Economic History of Wisconsin, 1860–70," in *Proceedings of the State Historical Society of Wisconsin*, 1907, p. 214.

[2] See, for example, the La Crosse *Republican and Leader*, January 20, 1872, and the *Wisconsin Lumberman*, 2:490 (August, 1874).

inspired with the ambition to keep in their own hands the manufacture of all the timber that had been going downstream. By 1870 the chief sawmilling centers of western Wisconsin had visions of supplanting St. Louis and Chicago as the great lumber marts of the country.[3]

Having accepted these hypotheses, the lumbermen of the state bent every effort to bring about the construction of suitable railroad lines. They sold wood and lumber to railway companies at reduced prices, induced local governments to vote financial aid, and bombarded the state legislature and Congress with proposals for subsidies to prospective railroads. Many lumbermen and pine owners subscribed personally for railway stock and bonds or built the necessary sidetracks in their towns. Some railway projects were organized and wholly financed by interested lumbermen, a few of whom even tried to entice Eastern capital into Western railroad construction.[4]

The promoters of such railway schemes suffered a number of disappointments, but before the close of the seventies an efficient network of rails connected the pineries with their markets. The foundation for the first lumber-carrying railway in Wisconsin had been laid in 1860 with the extension of the Chicago and North Western road from Oshkosh to Green Bay. When in 1871, at the insistence of lumbermen and with material aid from them, it was extended from Green Bay to the Menominee River, the entire eastern pinery had an outlet to Western markets that eliminated the necessity of shipping to Chicago. In 1877 the Wisconsin Central, the most important of the lumber roads, completed a line from Milwaukee to Ashland with branches to

[3] *La Crosse Leader,* July 23, 1870; *Eau Claire Free Press,* March 2, 1871, April 11, 1872, November 1, 1883; *Wisconsin Lumberman,* 1:3 (October, 1873); 2:46, 490 (April, August, 1874); 3:46 (October, 1874); Hotchkiss, *Lumber and Forest Industry,* 380.

[4] Memorandum of an Agreement for the sale of ties to the Mineral Point Railroad between Moses M. Strong and Albert C. Pinney, January 28, 1854, in the Strong Lumber Papers; S. S. Kepler to Orrin H. Ingram, March 3, 1870, in the Ingram Papers; W. A. Holt to the Holt Lumber Company, June 14, 1889, in the Holt Lumber Company Papers; *Eau Claire Free Press,* October 31, 1867, October 3, 1872, January 23, 1879; *Chippewa Union and Times,* March 6, 1869; *Chippewa Herald,* August 30, 1878, September 5, 26, 1879, July 7, October 13, 1882; Oshkosh *Northwestern,* May 10, 1883; Usher, *Wisconsin,* 1:207.

WHi(D489)9624

Barron County, Wisconsin, 1914. Lombard steam log hauler made by Phoenix Mfg. Co., Eau Claire, Wisconsin, on the Rice Lake lumber camp overland spur. As the photo shows, this machine could pull tremendous trains of logs loaded on sleds.

WHi(X3)18354

Otter Creek, Wisconsin, 1907. Water tank sled used for icing logging roads by the Chippewa Lumber and Boom Company operations. Iced roads, especially grooved ones, made hauling of loads of logs much easier. Photo probably by Charland.

Green Bay and Portage. Sawmills promptly sprang up in the wake of construction. Twelve miles west of Stevens Point the Wisconsin Valley Railroad, starting at Tomah, intersected the Wisconsin Central and stretched onward to Wausau. By 1874 the Green Bay and Minnesota Railroad was running from the Bay to the Mississippi, crossing the Wisconsin River at Grand Rapids and connecting with the La Crosse and Trempealeau, the Wisconsin Valley, the Wisconsin Central, and the West Wisconsin lines.[5]

In 1879 an arrangement was perfected whereby the Chicago, St. Paul, Minneapolis and Omaha, the St. Paul and Sioux City, the Sioux City and Pacific, and the Kansas City, St. Joseph and Council Bluffs railroads agreed to cooperate in carrying lumber over their lines at rates that enabled Wisconsin lumbermen to compete with Chicago and older distributing centers for the Western market. This cooperative network, called the "Wisconsin Lumber Line," approached perfection two years later when the officers of the Wisconsin Central agreed to extend to Mannville the shipping territory subject to the agreement. Before 1890 other important links in the network had been supplied by the construction of the Milwaukee, Lake Shore and Western line from Reedsville to Rhinelander, the St. Paul, Minneapolis and Sault Ste. Marie road, a line up the Chippewa valley from the mouth of the river, and a road connecting the St. Croix River with Lake Superior.[6]

The building of railways through the richest pine areas of the state effected sweeping changes in the logging and milling branches of the lumber industry. In the first place, they were a convenient means of getting lumberjacks to their camps; all logging railroads sold special tickets to the larger companies for this purpose. But they served an even more important function as carriers of logs to the mill. Immense tracts of pineland which had hitherto been almost worthless because they were too far

[5] Merk, *Economic History of Wisconsin during the Civil War Decade,* 80; Stephenson, *Recollections,* 174; *Wisconsin Lumberman,* 1:3, 188 (October, 1873; January, 1874).

[6] Oshkosh *Northwestern,* June 16, 1881; *Chippewa Herald,* December 5, 26, 1879, January 16, 30, February 20, March 12, 1880, September 1, 1882, January

from driving streams now came into great demand. Rough logging railroads were built from these timberlands to the main rail lines, sometimes by the logging companies, sometimes by the railway company itself in the expectation that settlers would replace the loggers on the cutover land. On main railroad lines, as formerly on watercourses, sawmilling towns began to spring up. The Chippewa River and Menomonie Logging Railroad, built about 1887 at a cost of over a hundred thousand dollars, was fairly typical. Its principal station was Verona, where it joined the Sault Ste. Marie. It had two locomotives, one for switching and another for hauling the cars; the average train comprised from forty to sixty cars, each carrying about twenty-five thousand board feet of logs. It made five trips daily from the woods to the main line.[7]

The establishment of direct rail connection between sawmills and retail lumber markets had a profound effect on the carpentry and building trades as well as on the lumber industry itself. Until 1879, when the Wisconsin Lumber Line brought a great new market within reach, the lumbermen of Wisconsin made little effort to distribute their product to other than wholesale markets on the Great Lakes and the Mississippi River. Moreover, they made virtually no attempt to dress their product for the use of the consumer. Shipping their lumber in rough form, they allowed the wholesale or retail dealer to reap the added profits to be derived from finishing it. After a wider area became accessible by rail, however, Wisconsin millowners began to make a bid for control of the entire market. Nearly every mill began to pile its lumber for seasoning before shipping it. As more millowners came to realize that it was remunerative to sell direct to the consumer, they added dry kilns for more rapid seasoning of the product and planing mills to make it ready for use.

18, 1884; La Crosse *Republican and Leader,* January 27, 1872; Hotchkiss, *Lumber and Forest Industry,* 441, 485, 517–521.

[7] Henry Pratt to Ingram and Kennedy, December 3, 1877, and James Barker to C. A. Chamberlin, October 16, 1885, in the Ingram Papers; La Crosse *Chronicle,* January 29, 1880, February 25, 1886; Oshkosh *Northwestern,* November 25, 1880; *Eau Claire Free Press,* February 5, 1880, January 18, July 19, 1883, November 5, 1885, February 27, 1890, January 12, December 21, 1893; *Chippewa Herald,* September 28, 1894, May 19, 1899; Jones and others, *Lincoln, Oneida, and Vilas Counties,* 142; Bruncken, *Forests and Forestry,* 80–81; *Wisconsin Lumberman,*

Soon lumber was being given an interior finish in factories owned by the sawmill or by independent operators in the milling centers. Some of the mills were now prepared to ship by rail a "house bill complete, from the rough sills and frame-work to the finished doors, casings, flooring and mouldings, and the elegant hardwood mantel or elaborately finished library." Thus the products of the factories came to supplant much of the handwork formerly done by the carpenter and builder. Still another result of the direct sale to consumer or retail yard was the development of a uniform system of grading.[8]

In western Wisconsin these developments were heralded as early as 1872, when the Knapp-Stout Company, largest in the state, withdrew almost entirely from the river and contracted with the West Wisconsin and Sioux City railways to carry its product to market. Others followed suit, but many were at first unwilling or unable to tie up capital in lumber piled for seasoning. Those who continued to raft their lumber, however, soon learned that competitors who used the railways were getting as much for their stock at the mill as downriver wholesalers paid at the market. By 1881 nearly two-thirds of the lumbermen on the Chippewa were piling their lumber and shipping it by rail, and Mississippi River millowners were beginning to build mills along the railroads in northern Wisconsin. Eau Claire was confident that it would soon be the principal market in the Northwest.[9]

The significance of these developments was not immediately apparent to the older marketing centers, for trade reports from the river towns at first belittled the importance of railways as lumber carriers. In 1874 St. Louis dealers were expressing satisfaction over the great expansion of the plains market which the railroads were achieving. It was not long, however, before the

1:300 (February, 1874); 2:9, 363 (April, July, 1874); 3:245 (December, 1874). The Holt Lumber Company Papers include a map of a typical logging line, the Thunder River Railroad. This road, built about 1909, cost the company $45,500.64 to construct and $4,494.59 to operate during the 1909–10 season.

[8] Hotchkiss, *Lumber and Forest Industry*, 380; George S. Long to Orrin H. Ingram, October 29, 1896, in the Ingram Papers.

[9] La Crosse *Republican and Leader*, July 20, 1872; *Eau Claire Free Press*, January 22, 1874, August 1, 1878, July 31, 1879, March 10, 1881, September 20, 1883; *Chippewa Herald*, October 1, 1880, February 25, 1881, September 1, 1882.

mills of western Wisconsin were being viewed as dangerous rivals by Chicago and Mississippi River wholesalers. From Chicago came a dispatch in 1883 which revealed that lumber merchants there were "watching the movements and intentions of the Eau Claire operators" and viewing with "a feeling kindred to positive alarm" their various moves to attract the Western market. The fears of the Chicago dealers and the ambitions of Wisconsin millowners were to result in a bitter conflict during the 1890's, but for the time being Chicago had no real cause to fear the lumbermen of western Wisconsin. The refusal of some of the railroads to give a favorable rate to Eau Claire and other Wisconsin millowners had by 1886 actually persuaded many lumbermen to return to rafting.[10]

A more serious immediate threat to the supremacy of the Chicago wholesale trade came from the mills of eastern Wisconsin and western Michigan, the source of its supply. The temporary economic advantage of railroad over lake transportation had induced these millowners also to ship orders directly to consumers and retailers. By 1888 the bulk of the product from the Green Bay region was being carried by rail. Even Chicago dealers were beginning to order shipment by rail, for there were definite advantages in doing so. Railroad cars could move in winter when the lake boats were tied up. The dealer to whom lumber was sent by rail could order exactly what he wanted, whereas shipment by lake boat meant that he had to unbalance his stock by taking a whole cargo, which, because of the necessity of filling the barges, always contained some lumber he did not need. Railroad transportation also permitted him to transfer his yard from the boat dock to a site nearer the local consumer, who was often more important to him than the out-of-town wholesale trade. Finally, a wholesale dealer receiving his order by rail could eliminate the cost of handling at his yard by rebilling an entire car to a country retailer.[11]

[10] *Eau Claire Free Press,* November 1, 1883, September 24, 1885, May 13, 1886; *Chippewa Herald,* June 19, 1885, February 11, 1887.

[11] Tunell, *Lake Commerce,* 55; Stephenson, *Recollections,* 83; Hotchkiss, *Lumber and Forest Industry,* 291, 298; W. A. Holt to Holt Lumber Co., June 12, 1889, in the Holt Lumber Company Papers.

By 1890 the railroads were carrying more lumber out of Wisconsin than the rafting streams. The rafting of lumber never ceased entirely, however. Indeed it was likely to spread whenever rate differentials favored it, and on the Mississippi it was encouraged by the large size of the rafts, which reduced the unit cost. But on streams like the Wisconsin, where rafting was exceptionally dangerous, the railroads were a welcome innovation; there the lumber raft had virtually disappeared by the mid-eighties. During the seven years from 1863 to 1870 the North Western Road carried only some ninety-five million board feet of lumber to Chicago from the Green Bay and Wolf River pineries, exclusive of shingles, lath, and staves. In 1885 La Crosse and Onalaska were sending out by rail an average of more than two million board feet weekly. In 1882 the Lumber Line carried nearly three hundred and sixty-three million board feet, and in 1884 seven hundred and thirty-seven million feet. The Wisconsin Valley Road at the peak of production in 1892 carried to market more than four hundred and sixty-four million board feet of lumber.[12]

This enormous amount of lumber was shipped directly to markets all over the nation. As early as 1874 Stevens Point lumbermen were shipping by rail to Denver and Omaha, and orders continued to come in from the plains states all through the seventies and eighties. In 1891 and 1892 the Northwestern Lumber Company of Eau Claire sent fifty-two hundred carloads of lumber to twenty-three states and territories.[13] Direct shipments became so common after the mid-seventies that St. Louis declined

[12] Moses M. Strong to Directors of Nekoosa Lumbering Co., September 13, 1886, in the Strong Lumber Papers; *Eau Claire Free Press,* July 2, 1885, September 1, 1887; Oshkosh *Weekly Northwestern,* October 1, 1885; *Chippewa Herald,* July 29, 1887; Hotchkiss, *Lumber and Forest Industry,* 517; Merk, *Economic History of Wisconsin during the Civil War Decade,* 80.

[13] The shipping record of this company suggests how Wisconsin lumber was distributed throughout the nation. During 1891 and 1892 Illinois received 753 carloads; Missouri, 751; Minnesota, 716; Iowa, 595; South Dakota, 523; Kansas, 519; Wisconsin, 387; Nebraska, 331; Indiana, 67; Colorado, 53; Massachusetts, 40; Ohio, 25; Washington, D.C., 10; Kentucky, 7; Wyoming, 7; Rhode Island, 6; North Dakota, 5; Oklahoma Territory, 3; Pennsylvania, 3; Connecticut, 2; Montana, 2; Utah, 1. Another company in the same district shipped lumber in 1894 to New York, Delaware, Michigan, New Jersey, and Maine in addition to most of the places mentioned above. *Eau Claire Free Press,* April 27, 1893; *Chippewa*

as a lumber market, and Chicago remained the only serious competitor of the larger mill towns.[14]

The railroads not only served the lumber industry as carriers but were themselves one of its best markets, for they used enormous quantities of wood. In the year 1867, when they were still in an early stage of development, Wisconsin railways used an estimated hundred and eighty thousand cords of wood for fuel, fourteen hundred trees for sleepers per mile of roadbed (which had to be renewed every six years), and lumber for bridges, piles, turntables, depots, cars, and many other adjuncts of railroading. All this required the timber from one hundred and ninety-seven acres, or a strip "one quarter of a mile wide the entire length of every railroad in the state." By 1874 the railroads of the country were using about a million acres of timber every year, much of which came from Wisconsin. In 1899 railways placed orders for five million ties in northern Wisconsin. Lumber journals spoke truly in 1911 when they observed that the country's railroads were the lumber industry's best customers.[15]

Railroad transportation was not without its disadvantages, however. It tended, especially before 1880, to be more expensive than rafting and for this reason never entirely replaced rafting in most lumbering districts, even when the roads were able to lower their rates. Often, too, the lumbermen were unable to obtain enough cars during the rush season. Still other deprecations came from the conservation interests, which regretted the impetus that the railways gave to the destruction of the forests of Wisconsin.[16]

Herald, April 28, 1893; Recapitulation of Rail Shipment for 1894, in the Ingram Papers.

[14] *Eau Claire Free Press,* September 11, 1879, May 10, 1883; *Northwestern Lumberman,* October 21, 1876; *Wisconsin Lumberman,* 1:394 (March, 1874).

[15] Increase A. Lapham, Joseph G. Knapp, and Hans Crocker, *Report on the Disastrous Effects of the Destruction of Forest Trees, Now Going On So Rapidly in the State of Wisconsin* (Madison, 1867), 25; Railway Business Association, *The Lumber Trade and the Railways* (New York, 1911), 3–6; *Eau Claire Free Press,* December 7, 1899; La Crosse *Chronicle,* May 26, 1881; *Wisconsin Lumberman,* 2:174 (May, 1874).

[16] Oshkosh *Northwestern,* June 21, 1866; *Wisconsin Lumberman,* 1:119, 250 (December, 1873; January, 1874); *Eau Claire Free Press,* November 18, 1886, November 26, December 31, 1889; *Chippewa Herald,* February 10, 1888; W. A. Holt to Durr and Rugee, July 16, 1888, and George H. Holt to Holt Lumber Co.,

But the greatest irritation with the railroads was born of dissatisfaction with their rate-making policies. Having placed themselves at the mercy of the railroads by building up a sales organization dependent on them, many lumbermen of Wisconsin found their business menaced by the disposition of some roads to give favorable rate differentials to Chicago. As early as 1865 railway rates were operating in favor of Chicago and to the disadvantage of the Mississippi River towns that received their lumber from Wisconsin mills. In that year, for example, when it cost $13.45 to ship a thousand feet of lumber from Chicago to Grinnell, Iowa, a distance of more than three hundred miles, it cost $9.00 to ship it a third that distance from the river town of Mucatine, Iowa, to Grinnell. The reason for the lower rate from Chicago was that the railways which carried grain to the city from west of the Mississippi needed a load for the cars that must return for more grain. As a result Chicago began to supply with lumber a region that had formerly been the exclusive territory of the river centers of distribution. Dealers in these towns who wished to keep their customers west of Iowa City were forced to buy their lumber in Chicago.[17]

Wisconsin lumbermen thus had the same grievances against the railroads as did the Western farmers who in the late sixties and early seventies launched their own movement for governmental regulation of the common carriers. As early as 1865 Cadwallader C. Washburn, prominent Wisconsin lumberman and politician, had declared that railway combinations were increasing rates to the point where they were extortionate, and that it was the duty of the legislature to take steps at once "to protect the public from further combinations, consolidations, and extortions of different railroad companies of the State"; reasonable rates should be set by law, and the roads should be compelled

October 21, 1889, in the Holt Lumber Company Papers. See also Lapham and others, *Destruction of Forest Trees*, 25.

[17] R. Musser & Co. to Ingram and Kennedy, May 25, June 27, 1865, and W. H. Day to Orrin H. Ingram, June 17, 1869, in the Ingram Papers. This does not mean that all Chicago dealers viewed the general policies of railroad corporations with unalloyed pleasure; for examples of criticism, see W. A. Holt to the Holt Lumber Co., May 28, 1889, and George H. Holt to W. W. Rice, May 29, 1889, in the Holt Lumber Company Papers.

to handle freight "in the order it may be offered, and without favoritism of any kind." The freight-rate issue came to govern the lumberman's attitude toward many of the local candidates for the state legislature. In 1874 the legislative debate preceding the passage of the Potter Act revealed how strong was the movement for the regulation of railroads among legislators from the lumbering districts.[18]

State legislation failed to control the railroads, but the lumbermen had another and more promising recourse in their struggle for rate adjustments. Unlike the farmers, they could form organizations capable of negotiating successfully with the railroad companies. In this movement the millowners of western Wisconsin had the cooperation of Minnesota lumbermen, who in 1878 were paying $51.50 a carload for delivery at Sioux City, whereas Chicago dealers, nearly twice as far away, actually paid a dollar and a half less. In August, 1878, a representative group of lumbermen met in St. Paul with officials of the Sioux City and Western Wisconsin railroads and effected an agreement for the transportation of lumber from the northern tributaries of the Mississippi to Southern and Western markets on a through-rate basis.[19]

This arrangement worked temporarily to the advantage of the lumbermen, but within a few years the millowners of western Wisconsin again found themselves in difficulties with the carriers. In February, 1884, the general freight agents of Missouri River and Northwestern lines met in Chicago and decided to advance from ten cents to fifteen cents per hundred pounds the rate on lumber shipped from Chicago to the "Missouri River points": Omaha, Council Bluffs, Kansas City, St. Joseph, Atchison, and Leavenworth. Pending the decision of an arbitrator, they set a temporary differential of five cents above the Chicago rate on shipments from Chippewa Falls, Eau Claire, Fond du

[18] Oshkosh *Northwestern,* November 23, 1865; Alexander Meggett to Orrin H. Ingram, October 2, 1867, in the Ingram Papers; *Wisconsin Assembly Journal,* 1874, p. 635; *Wisconsin Senate Journal,* 1874, pp. 385, 423; *Wisconsin Legislative Manual* (Madison, 1874), 448 ff.

[19] *Eau Claire Free Press,* August 1, 15, 1878; Oshkosh *Northwestern,* August 8, 1878.

Lac, and Menomonie, six cents from Oshkosh, and eight and a half cents from Wausau and Stevens Point. These temporary rates Wisconsin lumbermen did not oppose, but a storm broke when the decision of the arbitrator was announced in May. The Bogue decision increased the differential on shipments from Menomonie, Eau Claire, and Chippewa Falls to six and a half cents and stipulated a one-cent differential from La Crosse.

Since Kansas City was about equidistant from Eau Claire and Chicago, whereas Omaha was actually closer to Eau Claire than to Chicago, Wisconsin lumbermen fumed over the imposition of so high a differential. All but two of the sawmills on the Chippewa reverted to rafting, and newspapers throughout the Northwest attacked the railroads and the Chicago lumber dealers for unjust discrimination. Finally, in August, the disagreement was adjusted by the restoration of the former differentials on shipments from Eau Claire and Chippewa Falls. There followed a lull in the rate controversy that enabled the Wisconsin lumbermen to enjoy to the full the prosperity sweeping the country as the business cycle, climbing from the depression of the mid-seventies, began to approach a new peak. By fall of 1885 the fight over rates had become, for the moment, "a thing of the past." The lumber industry was in so prosperous a condition that it overtaxed the carrying capacity of the railroads. Lumbermen would have been only too happy to pay high rates if they could have obtained the cars they needed to ship their orders. The railroads took full advantage of these conditions. Already in April the Bogue differentials had been re-established, and the Wisconsin lumbermen, momentarily blinded by prosperity, failed to make a serious protest.[20]

It was not long, however, before the Eau Claire millowners began to feel the disadvantageous effects of the restored differentials. In 1888 an Eau Claire newspaper estimated that discrimination against the Chippewa valley amounted to a dollar and thirty cents per thousand feet of lumber in favor of Chicago,

[20] *Eau Claire Free Press,* February 28, March 13, May 29, June 5, August 7, 14, 1884, April 2, October 29, 1885; *Chippewa Herald,* May 23, June 13, 1884; Oshkosh *Weekly Northwestern,* October 1, 1885.

a dollar and ten cents in favor of La Crosse and Winona, ninety cents in favor of St. Paul and Minneapolis, and twenty cents in favor of Oshkosh. This situation arose just after the passage of the Interstate Commerce Act in 1887 had provided a new weapon against discrimination. Before appealing to any governmental agency for relief, however, Eau Claire mill men sent a Citizens' Committee to the offices of the Chicago and North Western, the Chicago, St. Paul, Minneapolis and Omaha, and the Chicago, Milwaukee and St. Paul with one last plea that unfavorable differentials be adjusted voluntarily. When the petition was denied, western Wisconsin began again to advocate state regulation. At the same time the lumbermen of the Wisconsin valley also organized an association to work for more equitable freight rates. The stage was set for the entrance of federal authority.[21]

The impending appeal of the Chippewa valley lumbermen to the Interstate Commerce Commission was postponed for nearly a year after the railroads, having heard of the move, voluntarily promised a rate reduction. When the promise failed to materialize by spring of 1890, the lumbermen of Eau Claire and Chippewa Falls took up a subscription and proceeded with their case. They asked of the commission that the differential to Missouri River points be reduced to a half cent over the Chicago rate. For more than a year the commission took testimony and heard arguments. The Eau Claire Board of Trade, speaking for all local businessmen, alleged that the unfavorable differentials had cost the city more than four hundred thousand dollars in a single year. The railroad companies, on the other hand, argued that if the rate for the Chippewa valley were to be lowered, the manufacture of lumber at all river points farther south would disappear; in justice to them all rates must be reduced proportionately if the differential for the Chippewa valley were reduced. Interested lumbermen from La Crosse and Winona, perturbed over the prospect of too favorable rates for Eau Claire,

[21] J. M. Faithorn to Orrin H. Ingram, February 6, 1888, in the Ingram Papers; *Eau Claire Free Press*, January 12, July 12, August 16, 1888, January 25, 1889.

WHi(X3)18372

Holmes Sons Logging Railroad, probably about 1880–1890, ready to start a haul.

WHi(X3)11485

Thornapple River, Rusk County, Wisconsin, 1906. Log loader used by the John H. Kaiser Company. These large steam-powered machines greatly increased the speed with which railroad cars could be loaded with logs. Photo probably by Charland.

received permission to intervene in the proceedings in order to fight the Eau Claire case.[22]

After seemingly interminable delay the commission finally ruled, in June, 1892, that the rate from Eau Claire to Missouri River points should not exceed the rate from Winona and La Crosse by more than two cents when the rate from the latter points was eleven cents or less per hundred pounds, and by no more than two and a half cents when that base rate was not over sixteen cents. Complying with the order, the Chicago, Milwaukee and St. Paul road reduced its rate from twenty-one and a half to eighteen and a half cents, but the Chicago and North Western and other competing roads insisted that they must make a corresponding reduction for all river points if they lowered the rate from the Chippewa valley. This would of course have nullified the effect of the commission's decision, for it had been not the absolute but the relative rate that Chippewa lumbermen had opposed.[23]

A few months after the commission announced its decision it became clear to the railroads that no effort would be made to enforce it. On September 1 the Milwaukee road abolished the reduction it had made, claiming that if competing roads made reductions for other points, Eau Claire and Chippewa Falls would not benefit from the new rates. In January, 1897, however, for reasons that are not quite clear, it again reduced the Eau Claire differential. Minneapolis and other river towns now complained that the Chippewa valley had an unfair advantage, and the Omaha road threatened to reduce the freight tariffs

[22] *Annual Report of the Interstate Commerce Commission,* 1892 (52 Congress, 2 session, Senate Miscellaneous Document no. 15, serial 3065, Washington, 1892), 16; *Eau Claire Free Press,* March 20, October 30, November 20, 1890, February 12, 26, May 7, 21, June 11, 25, September 24, October 1, November 19, 1891, January 21, 1892; La Crosse *Weekly Chronicle,* July 16, 1891; *Chippewa Herald,* March 28, 1890.

[23] *Eau Claire Free Press,* February 25, March 17, April 28, June 30, July 7, 21, 28, 1892; *Report of the Interstate Commerce Commission,* 1892, pp. 17, 100. The commission based its decision upon the principle that "each community is entitled to the benefits arising from its location and natural conditions, and the exaction of charges . . . by which those benefits are neutralized . . . contravenes alike the provisions and the policy of the statutes."

from many points along its line in order to equalize the market. At last, in June, the Interstate Commerce Commission decided to hold another hearing to investigate the failure of the railroads to comply with its decision. While carrying on its investigation the commission suddenly found itself deprived of jurisdiction. The Supreme Court of the United States in the Maximum Freight Rate Case, unrelated to the Eau Claire litigation, decided that the commission had no authority to set rates.[24]

Thus Chippewa valley lumbermen had failed to get through governmental action the freight differentials they considered to be equitable. But even had the commission succeeded in enforcing its decision, it is doubtful whether the problem would have been solved. The lumbermen of the two embattled Chippewa River cities could have been satisfied only at the expense of inflaming other lumbering towns. Fundamentally the issue can be resolved into a conflict between two concepts of rate-fixing. The Chippewa valley, whose opportunities for economical production were unsurpassed, insisted that advantages of position must be allowed full play, whereas the railroads and certain sawmilling centers less advantageously situated argued that railroad rates ought to be fixed with a view to equalizing opportunities. Long before the Interstate Commerce Commission was created the lumbermen of the Green Bay region had demonstrated that there was only one satisfactory way to force a railroad to reduce its rates. During the 1870's they had established a ferry that transported loaded railway cars from Peshtigo to Chicago by way of Lake Michigan. Faced with the prospect of losing all their lumber trade in that area, two railway companies cut their rates nearly fifty per cent in order to compete with the ferry.[25]

Chicago's success in the struggle for transportation advantages was due in no small measure to the failure of the railroads to supply western Wisconsin with adequate facilities and to the inability of Mississippi River towns to obtain favorable rate

[24] *Eau Claire Free Press,* July 28, August 4, September 1, 22, 1892, March 14, 1895, January 14, May 20, June 10, 1897; *Chippewa Herald,* April 12, 1895, January 15, February 12, June 4, 1897; Interstate Commerce Commission *v.* Cincinnati, N. O. and T. P. R. Company, 167 U. S. 479 (1897).

[25] Stephenson, *Recollections,* 83.

differentials. In 1879, for example, many Wisconsin millowners still found it cheaper to raft their products as far as Hannibal or some other river point and from there ship to the plains states by rail. Meanwhile the lumber dealers of Iowa and Illinois river towns, suffering from discriminatory rates, began to think of moving their businesses to Wisconsin, where the railroad situation, unsatisfactory as it was, seemed to offer at least the hope of survival. This was one of the reasons why Frederick Weyerhaeuser and his associates along the Mississippi River organized a stock company in 1880 to build mills at Eau Claire and Minneapolis.[26]

[26] *Eau Claire Free Press,* December 5, 1878, September 4, 1879, March 11, 1880; La Crosse *Chronicle,* March 11, 1880.

7. The Problems of Management

LUMBERING was a complex business, and the Wisconsin entrepreneur faced many problems besides the processing and transportation of his product. These two branches of the industry could be, and most often were, entrusted to skilled logging, sawing, and rafting technicians, but the accumulation and handling of investment capital, the reduction of operating costs, the assumption and minimizing of risks, and the prediction of marketing and price conditions as controls of production volume were strictly managerial responsibilities. Since these were the functions most immediately related to profits, it is not surprising that they should usually have been assumed by the owners personally, even when they were corporately organized.

None of these managerial functions was more difficult or more essential to success under frontier conditions than the raising of working capital. Granted that before 1860 only a small investment was required to start a lumbering enterprise, the fact is that few industrial pioneers possessed the necessary resources themselves or were in a position to command adequate credit. Failures were numerous in the early days, and almost without exception the great "barons" of Wisconsin lumbering were the men who had capital and dependable credit resources when they entered the industry. Two examples will illustrate what extremes of financial manipulation were used to raise capital.

In 1846 young John H. Knapp of Fort Madison, Iowa, entered the lumber business at Menomonie, Wisconsin, with a thousand dollars inherited from his father. Having survived the initial financial shoals, the Menomonie firm of Knapp and Wilson faced bankruptcy during the depression following the panic of 1857. The difficulties, fortunately, were overcome with the aid of another business partly owned by Knapp, the banking firm of Knapp and Eaton at Fort Madison. In January, 1859, Knapp's bank borrowed $36,000 in New York and Pennsylvania, and in August of that year, with declared assets of $33,500 (a sum evidently related to the total of the Eastern loans), it obtained a charter as a branch of the State Bank of Iowa. The organization now had the power to issue its own notes. Early in 1860 Knapp began to use these state bank notes to cover the Wisconsin firm's liabilities. On January 23 he sent a thousand dollars "to the Mill." The next day he sent a draft for five hundred dollars in notes of his bank "to insure our mills," and on the same day dispatched to Philadelphia a similar draft to redeem a note signed by the sawmilling firm. In May Knapp traveled to Menomonie and left five hundred dollars in bank currency with the lumbering company. By such means the depression was weathered; eventually the firm developed into what was at the time probably the largest unit of its kind in the world.[1]

Less successful were the efforts of Moses M. Strong to develop the water-power facilities at Nekoosa into a great lumbering enterprise. Having bought in his own name the land around the suitable power sites, Strong procured an act of incorporation in 1859 and organized a company to which he sold the whole of the property for forty thousand dollars, taking payment in stock. As the sole stockholder he still owned all the property, but he was now able to raise capital by selling shares to outsiders. He soon aroused the interest of two other men by allowing them to acquire twenty thousand dollars worth of stock at seventy-five per cent of par in return for their personal notes. The three men

[1] Knapp Diary, January 14, 17, August 4, 5, 13, October 13, 1859, January 2, 23, 24, November 5, 1860. For a detailed discussion of the founding of this company, see below, pages 124–127.

then agreed upon a plan of improvement that would cost eighty thousand dollars to complete.

The company's cash box was still empty, but the plan provided that the two new stockholders should between them invest twenty-five thousand dollars in cash, for which they would receive stock at par, and that Strong should invest fifteen thousand dollars, or find someone else who would, in return for stock at par. Thus the company would come into possession of forty thousand dollars, and the stockholders expected that after this amount had been expended in improvements they would be in a position to borrow forty thousand more to complete the project. Unfortunately Strong's two associates could not pay their notes when they became due, and the entire undertaking fell through.[2]

Generally speaking, Eastern money, raised by promotional schemes or by simple newspaper advertising, constituted the economic foundation of the Wisconsin lumber industry until the Civil War, after which it was supplemented by the investments of successful businessmen in Chicago and the Mississippi River market towns. This foreign capital, invested directly or used as a credit base for buying supplies, served the industry throughout most of the century. Before 1900, however, the more prosperous lumbermen of Wisconsin had accumulated so much surplus capital that their problem was one of investment rather than acquisition of new capital for the industry. Some began to play the New York stock market on margin; others transferred their lumbering activities to the far West or the South; a few invested heavily in cutover lands in Wisconsin, hoping to sell them at a profit to agricultural settlers. The change in status from a business seeking capital to one able to supply capital marked the full development of the industry in Wisconsin.[3]

Almost without exception the earliest lumber companies in

[2] Moses M. Strong to Ellen S. E. Strong, November 13, 1859, and Stock Certificates, dated 1861, in the Strong Lumber Papers.

[3] Letters to Orrin H. Ingram from E. R. Leland, March 3, 1877, Hugh Bellas, April 20, 21, 27, 1900, James L. Gates, May 2, 1900, and J. B. White, January 26, 1901, in the Ingram Papers; *Chippewa Herald,* September 23, 1871, April 2, 1880; La Crosse *Chronicle,* April 8, 1880; Oshkosh *Weekly Northwestern,* March 15, 1888.

Wisconsin were owned by individuals or were organized as partnerships. This was a simple procedure which served adequately to put capital to work so long as the average maufacturing unit remained small. During and after the Civil War, however, when lumbering became a large-scale industry and more capital was needed to run a sawmilling business, the corporate form of organization became increasingly common. Even when the partners in a lumber company could manage to raise enough money among themselves to enlarge their plant as the changing situation demanded, they commonly formed a corporation and issued stock to the members of the old firm, in order to enjoy the advantages of limited liability and permanence. Of the one hundred and sixty-eight Wisconsin lumber companies operating in 1896, all but sixty-eight were corporations.[4]

The capital invested in Wisconsin lumbering increased to a peak of $45,642,002 in 1897, constituting in that year almost a fourth of the total industrial capital employed in the state. According to the federal census of 1890, 36.93 per cent of the investment in the Wisconsin industry represented land. Of the remainder, 13.21 per cent was invested in logging and 16.52 per cent in the mill plant, and the remaining 33.34 per cent represented live assets. Figures compiled by the state Bureau of Labor and Industrial Statistics in the mid-nineties showed that 37.41 per cent of the lumbering capital was invested in land, buildings, and fixtures; 10.05 per cent was allocated to machinery, implements, and tools; and the remaining 52.54 per cent represented cash and credit capital.[5]

[4] *Report of the Wisconsin Bureau of Labor and Industrial Statistics,* 1897–98, p. 559. A typical partnership agreement is that made by Irenus K. Hamilton, Woodman C. Hamilton, William S. Finley, and Alfred K. Hamilton, dated March 2, 1868, in the Alfred Kittredge Hamilton Papers. The Stock Book of the Eau Claire Lumber Company shows that the original partners in a preceding firm held all the shares. Eau Claire Lumber Company Papers. Not until 1875 was an outsider allowed to buy stock, and then it was done in order to get a working partner who could help handle the increasing volume of office business. *Eau Claire Free Press,* August 19, 1875.

[5] *Report on Manufacturing Industries in the United States at the Eleventh Census: 1890* (52 Congress, 1 session, House Miscellaneous Document no. 340, pt. 22, serial 3026, Washington, 1895), pt. 3, p. 596; *Report of the Wisconsin Bureau of Labor and Industrial Statistics,* 1895–96, p. 451; 1897–98, p. 561; 1899–1900, p. 415.

Two factors tended to reduce the rate of return on this investment. In the first place, no lumber mill ever operated at full capacity the year round. In 1896 the average Wisconsin mill was operating only about a hundred and ninety days, which was appreciably less than two-thirds capacity. In the second place, operating expenses were fairly high in relation to the average selling price of lumber. In 1873 a trade journal stated that of the average price of twelve dollars a thousand feet only two dollars was available for stumpage costs and profit after the expenses of manufacture and transportation had been met. The estimated cost of bringing lumber from the stump to market was $8.68 per thousand feet in 1880 and $7.37 in 1894. An Oconto firm figured in 1892 that it cost $4.03 a thousand feet to cut logs on the Oconto River, $1.32 to drive them to the mill, 20 cents for boom work, $1.24 for sawing, and 53 cents for sorting and piling, a total of $7.32. To this should be added 33 cents for mill repairs, 25 cents for insurance and taxes, and 25 cents for office expense and salaries, making a grand total of $8.15 per thousand feet. These figures applied only to the Oconto River; the average cost of all lumber in pile at the mill came to $12.25 per thousand feet, to which dressing and shipping to market added another 93 cents. The average selling price at this time was between fourteen and fifteen dollars per thousand feet.[6]

Nevertheless the rate of profit was ordinarily satisfactory. About 1865, when prices were high, Moses M. Strong estimated that it cost him forty-five thousand dollars to produce three million board feet of lumber and transport it to St. Louis, where he sold it for sixty thousand dollars. In 1897 the gross product of

[6] *Ibid.*, 1897–98, p. 623; *Wisconsin Lumberman,* 1:12 (October, 1873); La Crosse *Chronicle,* January 8, 1880; Statement of Cost, January 6, 1893, in the Holt Lumber Company Papers; Expense Sheet, January, 1884, in the Ingram Papers. In the mid-nineties an average of 60.57 per cent of the operating cost went for stock; .05 per cent for rent; 3.73 per cent for salaries; 24.41 per cent for wages; 3.69 per cent for taxes; 2.13 per cent for insurance; 2.55 per cent for freight; 1.35 per cent for repairs; and 1.52 per cent for other expenses. *Report of the Wisconsin Bureau of Labor and Industrial Statistics,* 1895–96, p. 450. On distribution of expenses, see also letters to Moses M. Strong from M. M. Cothren, September 9, 1857, and S. J. Carpenter, January 24, 1859, in the Strong Lumber Papers, and Karl W. Woodward, *The Valuation of American Timberlands* (New York, 1921), 39.

the entire industry in Wisconsin amounted to more than twenty per cent of the invested capital; and the profit, before the payment of certain minor expenses, to nearly ten per cent. In general, profits had to be at least as large as this to attract capital, for many factors tended to make lumbering a risky undertaking. Weather conditions, vitally affecting the success of logging and of transportation by water, could not be controlled. Labor was largely nomadic, and training new men was expensive. The unfamiliarity of most capitalists with lumbering techniques forced them to depend on hired experts. All in all, it was likely to take a long time to realize any sizable profit on an investment.[7]

The greatest risk faced by the lumbering capitalist was loss by fire. Nearly every issue of a newspaper in a lumbering region reported the burning of a sawmill or an area of standing timber. The flammability of wood and the close proximity of fire to every steam-operated mill made the hazard a constant one. During a dry season the slightest spark might set a blaze in the woods or lumberyard that would result in enormous losses. In 1875 the entire business district of Oshkosh was reduced to a "black and ruinous waste" by a conflagration that destroyed two and a half million dollars worth of property. The Peshtigo fire of 1871 burned millions of feet of standing timber, swept away whole villages, and took the lives of fifteen hundred people. When the growth of the industry forced lumbermen to build larger mills, their dread of fire became so much the greater. In 1882 two mills near Eau Claire burned down, each with a loss of forty to fifty thousand dollars in excess of the insurance. Between 1885 and 1891 fire destroyed sawmills in Wisconsin valued at approximately $2,037,000. This was nearly fifty per cent of the fire loss of all the industries in the state.[8]

[7] *Report of the Wisconsin Bureau of Labor and Industrial Statistics*, 1897–98, p. 695; Statement of Cost [about 1865], in the Strong Lumber Papers; *Eau Claire Free Press*, December 10, 1885; Woodward, *Valuation of Timberlands*, 197–198.

[8] Oshkosh *Northwestern*, June 22, August 10, 17, 1860, October 27, 1861, May 18, 1865, May 13, 1875, August 9, 1877, August 11, 1894; *Chippewa Herald*, June 8, 1877; La Crosse *Republican and Leader*, June 17, 1877; *Eau Claire Free Press*, July 5, 1883, May 28, 1885; *Northwestern Lumberman*, January 5, 1878; *Report of the Wisconsin Bureau of Labor and Industrial Statistics*, 1888–89, p. 165; 1891–92, p. 113; 1893–94, pp. 116–117.

Because the hazard was so great, lumbermen often had difficulty obtaining adequate insurance against fire. Before the Civil War many insurance companies refused to insure sawmills at all, and most companies charged a high rate at all times. In 1857, for example, the Wisconsin State Insurance Company's premium for a wooden steam sawmill was nearly twice as much as for a wooden flour mill. To meet the needs of the lumbermen some insurance companies specialized in the protection of sawmill property. In 1853 the Wisconsin Lumbermen's Fire and Marine Insurance Company was underwriting losses incurred from fire and shipping disasters. Before the close of the century the more prominent lumbermen had organized their own insurance companies. In 1893, for example, several large white pine companies, most of them connected with the Weyerhaeuser interests, formed a mutual association for insuring one another up to twenty-five thousand dollars. Patterned after the famous British organization, it called itself the Lumbermen's Lloyd's of Eau Claire, Wisconsin.[9]

The installation of up-to-date machinery often entailed, besides the cost of the equipment itself, the payment of fees to the owners of various patents. In 1880 the lumbermen of Oshkosh, irked by the attempt of a Milwaukee firm to collect royalties on a device they had been using in their mills, formed a protective association to fight the claim. Three years later, however, the courts upheld the patent owner.[10]

Most lumbermen increased their incomes and the stability of their business by operating general stores and engaging in

[9] J. A. Noonan to Moses M. Strong, October 9, 1848, and W. B. Robbins to Julius T. Clarke, November 12, 1851, in the Strong Lumber Papers; *Tariff of Rates for Insurance against Fire, with Rules and Regulations for the Agents of the Wisconsin State Insurance Co.* (Chicago, 1857), 15–17; La Crosse *Republican and Leader,* October 9, 1875; *Charter of the Wisconsin Lumbermen's Fire and Marine Insurance Company, Approved March 22, 1853* (St. Louis, 1853); La Crosse *Weekly Chronicle,* January 22, 1891; *Eau Claire Free Press,* October 5, 1893; *American Lumbermen: The Personal History and Public Business Achievements of One Hundred Eminent Lumbermen of the United States* (3 vols., published by American Lumberman, Inc., Chicago, 1905–06), 1:309.

[10] Cyrus Brown to Ingram and Kennedy, May 21, 1865, in the Ingram Papers; Oshkosh *Northwestern,* August 19, September 9, 30, 1880, February 3, 1881, May 10, 1883.

other activities allied to lumbering. In 1873 a wholesaler advised an Eau Claire lumber firm that a company store of moderate size should produce an annual net return of about eight thousand dollars. Many of the stores of the larger companies carried an extraordinarily complete line of merchandise. In 1872 a Chippewa Falls company advertised clothing, carpets, woolen goods, dry goods, furs, ladies' winter "ware," laces, notions, boots and shoes, crockery, hardware, groceries, and meats. Early lumber companies also augmented their incomes by acting as sales or purchasing agents for Milwaukee, Chicago, and Eastern firms on a commission basis. Wisconsin lumber companies had many business and credit connections with Eastern firms.[11]

The most common by-product of a lumber mill, sawdust, could sometimes be made to pay income into the company treasury. In the early days it was usually treated as mere waste and dumped into the stream upon which the mill was located. This convenient procedure often disrupted navigation and created difficulties for milling companies downstream, sometimes actually creating new islands in the rivers. In the face of these problems and of legislation prohibiting impediments to navigation, the mills began experimenting with new uses for sawdust. A certain amount of it could always be sold as packing material. In 1860 it was discovered that hardwood sawdust was an effective fertilizer, and twenty years later sawdust alone was being used successfully to grow potatoes. Some farmers were induced to use sawdust for mulching purposes. Efforts were made to produce paper from it. But no really profitable utilization of sawdust, other than as a fuel, was discovered before 1900.[12]

The uncertain elements in the lumbering enterprise and the factors that tended to hold down profits placed lumbermen in an extremely delicate position when their markets were dis-

[11] Cephus Shekell to Moses M. Strong, October 4, 1850, in the Strong Lumber Papers; Abel Wright to Dole, Ingram and Kennedy, February 13, 1858, and A. T. Kerry to Orrin H. Ingram, March 4, 1873, in the Ingram Papers; Knapp Diary, March, 1860; *Chippewa Falls Democrat,* October 31, 1872; *Chippewa Herald,* December 17, 1875; *Eau Claire Free Press,* December 23, 1875.

[12] *Chippewa Herald,* May 5, 1899; *Northwestern Lumberman,* March 2, 1878; Oshkosh *Northwestern,* August 24, 1860; La Crosse *Chronicle,* March 11, 1880; Hotchkiss, *Lumber and Forest Industry,* 294.

turbed by a financial panic. In 1857 and 1858 many found themselves wholly unable to make any collections or to meet their own obligations. During the depression following the panic of 1873 the Union Lumber Company of Chippewa Falls failed, jeopardizing the welfare of the entire community, and many concerns throughout the state accompanied it into receivership. Succeeding panics had an equally adverse effect upon the industry. An examination of market conditions illustrates the sensitivity of the lumber industry to general price fluctuations.[13]

When the prairie regions of the Middle West were first being settled and little attempt had yet been made to exploit the forests of the Northwest, the demand for building materials so far outran the supply that lumber brought exceedingly high prices at almost any market west of the Appalachians. Along the Mississippi in 1830 Wisconsin River lumber brought from fifty to sixty dollars a thousand feet. Within a few years, however, as means were developed for distributing the pine timber of Michigan, Wisconsin, and Minnesota, the supply was brought into better accord with demand. By 1839 sawed pine could be bought at Mineral Point for about thirty dollars a thousand feet, and ten years later lumber prices had dropped to the norm about which they were to fluctuate for the rest of the century. In 1849 river and lake markets were paying ten to twelve dollars for rough lumber.[14]

The early fifties witnessed one of the booms that have always preceded panic and depression in this country. Prosperity was general in the West; farmers and manufacturers were enjoying

[13] A. M. Dole to Orrin H. Ingram, May 30, 1857, in the Ingram Papers; Hotchkiss, *Lumber and Forest Industry,* 294; *Chippewa Herald,* September 10, November 26, 1875, May 12, 1876, December 21, 1877, March 12, 1878, August 8, December 26, 1879, November 21, 1884, February 23, 1900; La Crosse *Republican and Leader,* October 4, 1873, April 17, 1880; *Northwestern Lumberman,* January 25, February 22, August 9, 1879; La Crosse *Chronicle,* April 22, October 28, 1880. In the following account of prices, no reference is made to the various grades of lumber, since grading was not standardized in the Midwest until after 1890. The prices given are average for mill-run lumber.

[14] Moses M. Strong to McLean and Stillman, May 2, 1839, and John Catlin to M. M. Strong, May 13, 1849, in the Strong Lumber Papers; Knapp Diary, 1848–49; John T. Kingston, "Early Exploration and Settlement of Juneau County," in *Wisconsin Historical Collections,* 8:401 (Madison, 1879); Ellis, "Upper Wisconsin Country," *ibid.,* 3:443 (1857).

high prices and a good market. The lumbermen, reaping the profits of heavy sales to their prosperous city and farm customers, were unable to meet the demand. In 1856 the price of lumber soared as high as thirty dollars, and throughout the early months of 1857 it never dropped below sixteen dollars. By autumn, however, the favorable market had come to an abrupt end. Within a few months the price fell to eight dollars, and during 1858 it dropped to new lows of four and five dollars a thousand. For three years a pall of gloom hung over the lumber mills. Many concerns that had borrowed freely to expand their operations in the happy days before August, 1857, were faced in 1858 with adamantine creditors and an overstock of sawed lumber for which there was no market. Bankruptcies multiplied. The firms that survived did so only by restricting credit operations to a minimum and practising rigid economy. Lumber trading came almost to a standstill, since wholesale and retail yards could buy only on credit.[15]

At the beginning of 1860 the situation was beginning to look more hopeful. An excellent grain crop enabled the farmers to buy lumber for the improvements that had been postponed for nearly three years. Prices again rose to about twelve dollars. But the revival was short-lived. The election of Abraham Lincoln to the presidency in November foreshadowed the secession of the South and civil war. To the lumbermen of Wisconsin this meant that the lower Mississippi, one of their best markets, would be cut off. The newly reborn optimism of Western businessmen again languished, and by 1861 the lumber industry was once more plunged into depression.[16]

When the war did break out, the business of the Middle West was for a short time nearly prostrated. Farmers who had held their grain for a higher price found in July, 1861, that it would

[15] A. M. Dole to Orrin H. Ingram, August 29, 1857, Dole to Ingram and Kennedy, January 8, 25, 1858, and James Gilbert to Dole, Ingram and Kennedy, March 12, 1858, in the Ingram Papers; *La Crosse Independent Republican,* May 19, 1858; Ellis, "Upper Wisconsin Country," in *Wisconsin Historical Collections,* 3:443; Ingram, *Letters of a Pioneer,* 24; *Eau Claire Free Press,* November 15, 1883.

[16] A. M. Dole to Ingram and Kennedy, February 17, 1860, in the Ingram Papers; Knapp Diary, September, 1860; Merk, *Economic History of Wisconsin during the Civil War Decade,* 61–63.

bring only thirty-five cents a bushel and their corn only ten cents. A lumber salesman reported to his firm that there was "nothing doing below Muscatine. They have no money and the farmers are not buying." In Missouri, where the division of political opinion had produced great instability, the lumber trade ceased altogether. By October it had become exceedingly difficult to make collections because of the derangement of the state's bank currency, and in December it became virtually an impossibility when the Eastern banks suspended specie payment. Moreover, the threat of civil disorder, which would exist until one or the other side should prevail, discouraged anyone from risking capital in Missouri.[17]

Before the close of 1861 the federal government's orders for lumber began to bring the Mississippi River trade out of the doldrums. The low price of the preceding summer, six dollars a thousand feet, gradually gave way to increases. During the next year conditions became appreciably better, although the farmers, plagued now by the prospect of increasing taxes, were still being cautious. In Missouri the disordered conditions of 1861 had disappeared by the following March, and before 1865 the wartime demand for lumber and the inflation of the currency had restored prosperity. In 1864 a large volume of Wisconsin lumber was sold at prices ranging up to thirty dollars a thousand feet. One small yard on the Mississippi made a profit of about seventeen hundred dollars in less than three weeks. When wheat advanced to a dollar and a quarter a bushel, many farmers started buying building supplies again, though others were disposed to wait for a drop in price.[18]

With the end of the war lumber prices again declined rapidly. Although the Southern market became a strong enough factor in the early months of 1865 to sustain prices at about twenty

[17] Letters to Dole, Ingram and Kennedy from A. Kempt, July 4, 1861, E. H. Playter, October 15, 22, 1861, and W. H. Padon, December 31, 1861, in the Ingram Papers.

[18] Letters to Dole, Ingram and Kennedy from E. H. Playter, October 9, 1861, and Gill and Brother, March 6, 1862; letters to Ingram and Kennedy from R. Musser & Co., February 10, 1862, December 29, 1864, W. V. Gill, January 18, 1864, and J. B. Randall, September 16, 19, October 19, 26, November 2, 1864, in the Ingram Papers; Stephenson, *Recollections,* 161.

dollars for a short time, the cessation of government orders and the decline of consumption brought steady reductions before the end of the year. By June the average price had dropped to fifteen dollars, and by December some lumber was bringing only ten dollars. During the next year the home demand was great enough to boost the price to nearly twenty dollars, but it was only temporary. By 1867 an agricultural depression had set in, and for three years thereafter the price of lumber fluctuated around twelve dollars.[19]

As the demand for lumber increased and the supply declined with every passing year, lumbermen saw the prospect of a constant advance in prices for the rest of the century. From 1870 until 1873 their expectations were fulfilled. Log shortages and heavy forest fires reduced the supply of lumber on the market at the same time that the demand was heightened by the great Chicago fire and the rapid settlement of the Western plains. Prices increased by two to four dollars. Early in 1873 many mills began to stock up for a heavy season, borrowing heavily to do so. The advancing prices of the past few years, the vast projects for developing the far West, and the seeming prosperity of the country convinced them that they were entering another boom period. The failure of Jay Cooke and Company and the panic it precipitated caught them wholly unprepared. Prices fell and markets disappeared. Creditors clamored for payment, but all the lumbermen had was a large supply of cut timber which they could not sell. Nearly all the mills shut down while their owners made a feverish effort to organize an association that would control production for the next few years. Despite their work, the oversupply of lumber continued to grow throughout the year 1874. In March, it was estimated, St. Louis had an overstock equivalent to three-fourths of its total sales in 1873. By October Chicago had seventy-five million board feet more on hand than it had had twelve months earlier. In the same month a trade jour-

[19] Telegrams to Orrin H. Ingram from R. Musser & Co. and J. H. Pierson, March 4, 1865, and letter from DeWitt C. Clark to Ingram and Kennedy, June 24, 1865, in the Ingram Papers; Hungerford and Brother to Moses M. Strong, August 8, 1867, in the Strong Lumber Papers; *La Crosse Weekly Democrat,* May 29, 1865; Oshkosh *Northwestern,* June 14, 1866.

nal estimated that Wisconsin mills had already produced enough to meet the demand for the rest of 1874 and all of 1875 unless business conditions improved. Throughout the year 1874 it was almost impossible to sell any lumber for cash or good commercial paper, and by 1876 prices had dropped to an average of about nine dollars.[20]

In 1877 lumber prices began to rise slowly. The lumbermen's organizations that had been established to control production took full advantage of a short log crop and a large grain crop to advance lumber prices as much as a dollar and a half a thousand during July, 1877. The slow advance continued through 1879, when some Mississippi River points were able to get as much as thirteen dollars a thousand. Extensive floods in the spring of 1881 helped to boost the price still further. They dissipated much of the supply of logs and cut timber on the Mississippi and so damaged the Chicago railways that they had to suspend service for a time; they also damaged farm buildings and other structures, which had to be repaired or replaced. Before June the best grades of lumber were selling for over twenty dollars. But that high price was only temporary; the next year prices were again ranging around the normal ten to twelve dollars.[21]

Then followed again a period of expansion and prosperity. In 1883 the price of lumber increased from the norm of twelve dollars to fourteen. After a slight retardation in the next two years it continued to advance in 1886. In that year forest fires and drought fortified the efforts of the lumbermen's associations to raise prices, and in the ensuing years the agricultural West bought large quantities of building materials, driving up the price to an average of about fifteen dollars in 1888. The demand continued to grow until the panic of 1893 reproduced the situa-

[20] *La Crosse Leader,* February 19, 1870; La Crosse *Republican and Leader,* November 4, 1871; *Chippewa Herald,* May 10, October 17, December 19, 1873, October 23, 1874; *Eau Claire Free Press,* January 22, 1874; *Wisconsin Lumberman,* 1:2, 94, 370 (October, November, 1873; March, 1874); 2:457 (August, 1874); 3:2 (October, 1874); Fernow, "American Lumber," in Depew, *American Commerce,* 1:201.

[21] *Eau Claire Free Press,* July 5, 26, 1877, February 3, May 19, 1881, November 23, 1882; *Chippewa Herald,* July 6, 1877, September 19, 1879; *Northwestern Lumberman,* March 15, April 12, October 25, 1879; La Crosse *Republican and Leader,* September 13, 1879.

tion of 1857 and 1873. This time, however, the recession did not last long. By 1895 the Mississippi Valley Lumbermen's Association was planning to increase prices. In 1896, when William McKinley and "sound money" won the presidential election, prosperity again returned to the Wisconsin industry, to continue into the twentieth century.[22]

Early conservationists believed that the rapid depletion of Wisconsin's forest resources would produce a progressive increase in the price of lumber. As early as 1867 an investigating committee reported that "a poor man cannot to-day build his wooden house for the same amount of money, that ten years since he could have built the same, after the lumber had been carried from 400 to 600 miles by water and rail from that same pinery."[23] The generally accepted idea that lumber prices were constantly increasing is not, however, borne out by the figures. The fact is that most of the time from 1860 to the close of the century they fluctuated very closely around a norm of ten to twelve dollars, despite the fact that the open-market price of stumpage increased tremendously. This increase was offset in the price of sawed lumber by competition among the smaller mills, better transportation facilities, the improvement of tools and machinery, more systematic business methods, and the mill-owners' failure to realize just how fast their timber supply was disappearing. Until after 1900 economies of organization operated to keep consumer prices fairly low.[24]

Nevertheless the conviction that the destruction of the American forests would raise the price of lumber to prohibitive heights served to unite the consumers of the Middle West with the conservation interests in a fight to admit Canadian lumber duty-free, as it had been admitted for a period before the Civil War. In March, 1855, a treaty had gone into effect which provided for reciprocally free admission of timber and certain of its products

[22] *Eau Claire Free Press,* January 18, September 27, 1883, August 27, 1885, August 26, 1886, March 30, 1893, June 13, 1895; La Crosse *Weekly Chronicle,* March 22, 1888; *Chippewa Herald,* July 27, 1888, August 11, 1893, November 13, 1896, August 13, 1897, May 19, August 11, 1899; Price Lists, 1888–1891, in the Holt Lumber Company Papers.

[23] Lapham and others, *Destruction of Forest Trees,* 28–29.

[24] Fernow, "American Lumber," in Depew, *American Commerce,* 1:201.

into Canada and the United States for a period of ten years and as long thereafter as neither signatory exercised its right to terminate the agreement on a year's notice. During the eleven years of its operation an increasing amount of Canadian lumber and timber found its way into the United States, but it was never enough to have much effect on the American industry. At the same time Canada consumed an increasing quantity of American wood products; in 1865 their value practically equaled that of American imports of wood products from Canada. Nevertheless Eastern lumbermen persuaded the administration to give notice of termination in January, 1865, and in March of the next year the treaty became inoperative. A tariff was immediately imposed on Canadian lumber.[25]

Wisconsin lumbermen had evinced little interest in this treaty. During the years it was in operation their business was in process of being established and their chief market was in the Middle West, where their transportation advantage, they thought, would always protect them against serious competition from Canada. Even after the Civil War, when American industrialists bestirred themselves to insure the continuance of the high wartime tariffs, the lumbermen of Wisconsin still seem to have been indifferent to protection for their own product. In 1869 Isaac L. Usher, a prominent Black River lumberman and a Democrat, declared that no articles should be taxed so high as to prohibit their importation. Natural laws, he said, would regulate commerce better than "discriminatory laws, passed through the influence of grasping capitalists."[26] It is probable that he was influenced more by the traditional tariff policy of his party than by any strong economic sentiments, but the very fact that he could argue against a tariff indicates that he did not believe that his industry would be helped by a high tariff nor harmed by a low one.

After 1870, however, several factors united to align Wisconsin

[25] Franklin B. Hough, *Report upon Forestry Prepared under the Direction of the Commissioner of Agriculture*, vol. 2 (46 Congress, 2 session, House Executive Document no. 37, serial 1922, Washington, 1880), 513; John Ise, *The United States Forest Policy* (New Haven, 1920), 94.

[26] *La Crosse Leader*, November 20, 1869.

lumbermen in favor of a high protective tariff on lumber. The development of marketing by rail not only tended to destroy much of the transportation advantage that Wisconsin had formerly enjoyed over Canada but created a nationwide market for its lumber, which came into competition with Canadian lumber at several points. Moreover, after 1870 the tariff became increasingly important as an issue between the two great political parties of the country. Most Wisconsin lumbermen were Republicans, either because of the economic or emotional situation in which they were placed by the Civil War or because they were natural heirs of the Whig tradition. Support of a high tariff on lumber was not wholly a matter of party allegiance, however, for by 1890 certain Democratic lumbermen were also opposing any reduction in the rates on lumber.

Consumer interests in the chief marketing centers soon became dissatisfied with the tariff on lumber. In 1870 memorialists from New England and New York petitioned Congress to remove the duty because it was a tax on the American consumer which returned only a small revenue to the government, and because they did "not deem it good policy to encourage the present waste and diminution of home-growth lumber from our fast waning forests by the imposition of said duties." Lumber interests countered with the argument that the profits in their industry were modest, even in Wisconsin, and that free lumber from Canada would so reduce the domestic price as to force them to "resort to the old wasteful ways of cutting timber."[27]

The development of the railroads, which enabled lumbermen to reach out for a national market, had changed their attitude toward protection. In the face of the pressure from consumers and conservationists they were gradually forced to enter the fight to maintain a tariff on lumber. When in 1872 Congress debated a bill for the removal of the duty on all materials that would be needed for the rebuilding of Chicago after its great fire, the *Chippewa Herald*, always vehement in its defense of the

[27] *Letter to the Committee of Ways and Means on Tariff on Lumber, 1870* (41 Congress, 2 session, House Miscellaneous Document no. 96, serial 1433, Washington, 1870).

lumbering interests, interpreted it as a "special drive" against the lumber industry of Wisconsin and Michigan. A few months later a congressman from Wisconsin tried to have lumber stricken from the bill.[28]

At this time the duty on all Canadian pine lumber in rough form was two dollars a thousand feet and somewhat higher on finished lumber. In 1874 President Grant asked the Senate to approve a new reciprocity treaty with Canada which would reduce these duties on lumber. Retail lumber dealers as a class supported the new treaty, but manufacturers throughout the country rose in opposition. A Wisconsin trade journal reminded its readers that Canadian lumbermen did not have to invest money in timberland nor assume the risk of forest fires, for they were allowed to cut timber on government land if they paid for the stumpage. The National Association of Lumbermen, in which Wisconsin members held prominent offices, went on record as opposed to the proposed treaty, declaring that "its effect upon the business which we especially represent would be most disastrous." The lumber manufacturers of the country managed to bring enough pressure to bear upon the Senate to prevent ratification of the treaty.[29] The brief struggle had brought out certain differences in interest between various branches of the industry and between different localities, but it had also demonstrated that the lumbermen of Wisconsin were becoming more resolute in their support of the lumber tariff.

During the next decade a determined assault on the lumber tariff was made by wholesalers and retailers of lumber, consumers, and conservationists. In 1877 the Chicago Lumbering Exchange supported an abortive attempt to reduce the duty on lumber, and five years later the lumber-selling and conservation interests besought the new Tariff Commission to recommend reduction. The wholesalers of Chicago argued before the commission that the imminent exhaustion of the forests in Michigan, Wisconsin, and Minnesota would bring to an end the develop-

[28] *Chippewa Herald,* February 24, April 20, 1872.

[29] *Wisconsin Lumberman,* 1:15 (October, 1873); 2:347–349, 459 (July, August, 1874); 3:21 (October, 1874); La Crosse *Republican and Leader,* September 26, 1874; *Chippewa Herald,* June 12, 1874.

ment of the vast trans-Mississippi prairies, for without a supply of cheap building material they would "cease to present inducements" to the immigrant and the capitalist. They pointed to the fact that the federal government had virtually donated the timber on its lands to the lumbermen by selling the lands at extremely low prices, whereas Canadian lumbermen had to pay the current market price for their stumpage. The free admission of Canadian lumber would postpone the exhaustion of the American supply, it would stimulate the settlement of the West, and it would not harm the American lumbermen. The commission recommended considerable reductions in the lumber duties, but Congress heeded the lumbermen and in 1883 voted to retain the old schedule.[30]

For the moment the menace to protection had been staved off, but the interests promoting free lumber continued to exert themselves, forcing the producers to remain alert. In 1884 Senator Philetus Sawyer, one of the most successful lumbermen of Wisconsin, announced that although he had no objection to the repeal of the lumber duty he had voted to retain it. In an effort to defend him an Oshkosh newspaper explained that free lumber would affect the sale of the Michigan product in the Northwest and that Sawyer's interests depended on successful competition with Michigan lumber. In the same year Congressman William T. Price, another Wisconsin lumberman, painted a dire picture of the results that could be expected from the removal of the lumber tariff. There were, he said, some hundred and five thousand persons in Wisconsin dependent on the lumber industry, all consumers of agricultural products. Competition with Canada would ruin the Wisconsin industry and force all these people into agriculture, thus increasing the supply of farm products and reducing the demand at the same time. If this principle

[30] *Argument for Free Lumber,* submitted to the special Tariff Commission of 1882 by the "wholesale lumbermen of Chicago" (16-page pamphlet, Chicago, 1882). Also printed in *Report of the Tariff Commission Appointed . . . 1882,* 2:1599–1603 (47 Congress, 2 session, House Miscellaneous Documents, vol. 3, serial 2117). See also *Forest Protection and the Tariff on Lumber: Spirit of the Press* (36-page pamphlet, New York, 1883), a compilation of opinions expressed by leading Eastern newspapers; Ise, *United States Forest Policy,* 94–95; *Chippewa Herald,* March 23, 1877; and *Eau Claire Free Press,* January 4, February 22, 1883.

were applied to all the states and to all the creative industries, he said, the results would be "astonishing."[31]

Despite the growing protectionist sentiment among lumbermen, the McKinley Tariff of 1890 reduced the lumber duty by one dollar, and in 1892 the House of Representatives again considered a measure to put lumber on the free list. The sponsor of this bill, William Jennings Bryan, claimed that the lumber tariff was an unjust tax on his Nebraska constituents and that his proposal was favored by a number of representatives from other lumber-consuming areas. The situation appeared really ominous to the lumber interests when agents of the Canadian government were invited to Washington to discuss reciprocity. Greatly alarmed, the lumbermen of the upper Mississippi Valley sent seventy of their number to the capital to protest the Bryan bill. They were joined by delegations from Southern and Pacific coast lumbering regions, and together the lobbyists managed to prevent the passage of the bill. By this time the pressure for free lumber from consuming districts had converted virtually all lumbermen in Wisconsin to protectionism. As early as 1888 Senator Sawyer had given up the attempt to straddle the issue, and now even so staunch a Democrat as Orrin H. Ingram of Eau Claire could refer to reduction of the lumber duty as a "great piece of folly."[32]

The attempt of the Cleveland administration to bring about a general reduction of the tariff in 1894 incited Wisconsin lumbermen to the most strenuous opposition they had yet put forth. The tariff of 1890 had decreased the duty on rough boards to a dollar per thousand feet. With this the lumbermen were far from satisfied, but it was better than the new Wilson bill, which would admit rough lumber duty-free, and they held protest

[31] Oshkosh *Northwestern,* January 17, 1884; *Eau Claire Free Press,* May 15, 1884.

[32] *Journal of the House of Representatives,* 52 Congress, 1 session (1891–92, serial 2919), 66; *United States Statutes at Large,* 26:582; *Chippewa Herald,* October 5, 1888, February 12, 1892; *Eau Claire Free Press,* February 11, 1892; La Crosse *Weekly Chronicle,* April 21, 1892; *American Markets for American Lumber: Effect of Reduction in Duty on Canadian Lumber upon American Trade* (24-page pamphlet, Chicago, 1892), 6. The Bryan bill was not reported out of committee.

meetings to make their feelings known. Taking note, congressmen from lumbering districts swung into action. Some of them distinguished themselves more for oratorical prowess than for logical argument. Congressman George B. Shaw, for example, producing statistics to prove that consumers could expect no benefit from a reduction of the tariff, showed that lumber prices had increased $1.68 between 1889 and 1892 despite the lower McKinley Tariff, and then went on to argue that removal of the duty would "strike down this great industry" by forcing lumber operators to engage in ruinous price wars. But such tactics were ineffective. The Wilson-Gorman Tariff, placing rough lumber on the free list, became law on August 27, 1894.[33]

But the lumbermen had just begun to fight. Newspapers in the lumbering regions of Wisconsin kept up a running fire of adverse comment on the effects of the new tariff. In 1895 they related how a Michigan stave factory had been forced to shut down because it could not compete with cheap Canadian labor. They announced that during the year ending June 30, 1895, more than six hundred million board feet of lumber had been imported from Canada, while "almost the same amount" was "lying around" in Wisconsin unsold. In the presidential election of 1896 they supported the protectionist, William McKinley, and marshaled all their forces against the Democratic candidate, William Jennings Bryan. Seventy-five thousand men, Chippewa valley newspapers told their readers, would be thrown out of work in Wisconsin and Minnesota lumbering districts if Bryan was elected. The Wilson-Gorman law admitted Canadian lumber free of duty while American men were idle and their families starving. "If Bryan is elected," said a Wisconsin lumberman, "our firm will not put in a log next year." But it was McKinley who was elected, and the lumbermen rejoiced.[34]

[33] *United States Statutes at Large,* 28: 521; *Eau Claire Free Press,* January 4, 25, February 1, 1894; *Chippewa Herald,* January 5, 1894; La Crosse *Weekly Chronicle,* March 29, April 5, 1894; [Arthur Hill], *When the Last White Pine Tree Is Gone* [Saginaw, Michigan, 1897], 10. The pamphlet is addressed "To the Finance Committee, United States Senate."

[34] *Eau Claire Free Press,* September 10, 1896; *Chippewa Herald,* January 25, 1895, October 9, 1896. The money issue in the campaign of 1896 also influenced the lumbermen, most of whom supported the gold standard.

Immediately after the election the lumbermen organized for an intensive campaign to raise the tariff on timber products at least to the level of the McKinley Act, under which the industry had enjoyed "its most prosperous period." In December a call went out for a lumbermen's convention at Cincinnati to discuss a tariff that would stop importations of Canadian lumber. Local associations in Wisconsin, meeting to select delegates to the Cincinnati convention, expressed their desire for a tariff of two dollars a thousand board feet, twice as high a duty as that of 1890. When Congress met in December, the lumbermen were ready with their arguments. Canadian lumbermen, they said, did not have to buy land in fee simple, they paid much lower taxes, and their laborers received lower wages. Thus their production costs were less. Moreover, the Canadian railways and the Great Lakes waterway put American manufacturers "at the mercy of Canadian competition." There was no danger that American timber would soon be exhausted nor that the imposition of a prohibitive duty would foster a lumber monopoly in the United States. Competition within the industry was "fierce"; combined with ample supplies, it insured the consumer a perpetually reasonable price. The importation of Canadian lumber did nothing to help save American forests. Since it was the coarser grades that were imported, at a price the American manufacturer could not meet, only the choicer grades of domestic lumber were marketable in the United States. The result was that the American producer took the best part of the tree and left the rest to rot in the forest. The United States government was losing at least a million and a quarter dollars annually by virtue of the removal of the lumber tariff, while the sale of stumpage in Canada increased. American lumbermen who favored reduction of the duty were all men with Canadian timber holdings, and "these Canadian interests are not entitled to consideration at the hands of an American legislature."[35]

Faced with this formidable array of arguments, the advocates

[35] *Chippewa Herald,* December 11, 1896; *Eau Claire Free Press,* December 17, 24, 1896; *Memorial of the National Lumber Convention to the Honorable Justin S. Morrill, Chairman, and Members of the Finance Committee, United States Senate* (Washington, 1897); [William S. Harvey], *Why We Want a $2.00 Duty on*

of tariff reduction replied rather ineffectually that American timber supplies would most certainly be exhausted soon unless foreign timber was drawn upon, that the lumber tariff had never paid any revenue to the American government, and that lumber could not be produced more cheaply in Canada than in the United States, as was proved by the fact that some Canadian firms bought American lumber and freighted it to Toronto for less than they could buy it at home.[36]

Their efforts were foredoomed to failure, for the issue was already decided. Wisconsin congressmen, having written their lumbering constituents for advice, received from them statements enumerating the advantages enjoyed by Canadian lumbermen under the free-trade system. Prominent millowners reminded Congress that the Wilson bill had removed the duty on logs and lumber on the theory that these were raw materials, whereas in actual fact they were not that at all. Sympathetic congressmen pointed out that a two-dollar duty would be "only" eighteen or twenty per cent ad valorem; many commodities were taxed much higher. The outcome was inevitable. The new Dingley Tariff gave the lumbermen what they wanted. Wisconsin millowners, elated over the double victory of protection and the gold standard, looked forward to a "great revival of trade."[37]

Lumber (16-page pamphlet, n. p., [1897?]), a letter addressed to J. C. Pritchard, United States Senate, "Replying to Arthur Hill's Argument."

[36] George W. Stevens, *Questions You Can't Answer as to a Duty on White Pine Lumber* [Washington, 1897?]; Hill, *When the Last White Pine Tree Is Gone.*

[37] M. Griffin to Orrin H. Ingram, December 15, 1896, in the Ingram Papers; *Chippewa Herald,* February 26, March 26, April 2, August 6, 1897; *Eau Claire Free Press,* July 15, 1897; *United States Statutes at Large,* 30:167.

8. The Trend Toward Monopoly

THE sensitivity of the lumber industry to price fluctuations made the idea of monopoly control, either by large-unit production or by combinations of units, exceptionally alluring to Wisconsin lumbermen after 1870. Although their industry was not particularly well suited to large-scale activities, lumbermen were seized by the "big business" concepts of the postwar years and steadily enlarged the scope of their manufacturing units. One by one most of the small companies, crushed by the expense of buying the modern machinery necessary for successful competition, were swallowed up or frozen out by larger concerns. The small company never disappeared entirely, but the character of lumbering in Wisconsin was henceforth to be set by the larger corporations.[1]

The growth of the individual mill is best revealed in the increase of its daily output. In 1848 a mill capable of cutting from nine to twenty thousand board feet a day was considered a large establishment. In 1865 a daily output of forty to fifty thousand board feet was remarkable, but only two years later the larger mills were capable of producing a hundred and fifty thousand board feet a day or even more. In 1873 a mill at Chippewa Falls

[1] Filibert Roth, *On the Forestry Conditions of Northern Wisconsin* (Wisconsin Geological and Natural History Survey, Bulletin no. 1, Economic Series no. 1,

averaged three hundred and twenty-five thousand board feet daily for twenty-five days.[2]

In 1874 an observer was struck "with the magnitude of the business being carried on" at the Union Lumbering Company's mill at Chippewa Falls. It was a three-story structure two hundred feet long and a hundred and eighty feet wide. On the first floor were its power works, on the second its rafting works, and on the third the ninety saws that cut the logs into boards. Three hundred men were required to keep the plant in action. In addition it operated shingle and lath works; and besides its mills it maintained three blacksmith and machine shops to keep its machinery in repair and a general merchandise store which in 1873 sold four hundred thousand dollars worth of goods.[3]

The urge to expand, which had been manifest even before the Civil War, carried lumbering companies into many strange fields of activity. In the early days of the industry it was not uncommon for a lumberman to engage in the fur trade. Many lumber companies ground flour for use in their camps or for sale in their stores, utilizing the power generated by their water wheels or operating a separate grist mill. Some operated steamboat lines and carried on an express business. In 1869 an Eau Claire lumberman was planning to put up a telegraph line from his city to Read's Landing. Almost all the larger lumber firms operated farms, and some lumbermen took a great interest in scientific agriculture; cashmere goats were introduced into northern Wisconsin by a lumber company. Other concerns operated factories for the manufacture of woolen cloth, wooden boxes, barrels, paper, and many other articles. In 1897 a firm on the Chippewa River owned a plant for the generation of elec-

Madison, 1898), 40; Cameron, *Governmental Forest Control,* 102; Merk, *Economic History of Wisconsin during the Civil War Decade,* 72–75. For a summary of the factors inimical to monopoly, see below, pages 139–140.

[2] Thomas I. Moorman to Moses M. Strong, April 23, 1848, in the Strong Lumber Papers; Oshkosh *Northwestern,* October 12, 1865, November 22, 1866; *Eau Claire Free Press,* July 11, 1867, December 6, 1883; *Chippewa Falls Democrat,* September 2, 1869; La Crosse *Republican and Leader,* June 29, July 6, 1872; *Chippewa Herald,* November 7, 1873, August 13, 1875; *Wisconsin Lumberman,* 1:201, 204–205; 2:600 (January, September, 1874).

[3] *Wisconsin Lumberman,* 1:204–205 (January, 1874).

tricity. Some lumbermen possessed patents of various kinds and conducted licensing businesses.[4]

The history of Knapp, Stout and Company of Menomonie embodies most of the typical features of the development of lumbering in Wisconsin, although the firm itself, because of its vast size, was by no means a typical one. It began to saw lumber on a small scale in 1846 when two pioneers, William Wilson and John H. Knapp, took over on the Red Cedar River a mill that antedated the Indian treaty of cession. Wilson, a former Pennsylvania lumberman, had engaged in contracting on public works, steamboating, and other activities after leaving his native state. By 1846 he was left with nothing but a collection of debts. Casting about for some means of making a living, he discovered the little mill on the site of what was later to become Menomonie. He realized immediately that its strategic position presented unusual opportunities for controlling the entire lumber industry on the Red Cedar and its tributaries, but he could find no one to advance him the money he needed to buy a share in its ownership.

His future partner was then a young man approaching maturity. Knapp's father, a veteran of the Black Hawk War, had in 1835 brought the ten-year-old John and the rest of his family from their native New York to Fort Madison, Iowa. Two years later he died, apparently leaving his family sufficient means of support. In 1845 young Knapp entered law school at New Haven, Connecticut, but remained only a year. Having now reached the age of twenty-one, he received an inheritance of a thousand dollars and, as a man of capital, was sought out by Wilson, who finally convinced him that his future could be best developed at Menomonie. In 1846 Knapp and Wilson pur-

[4] Samuel H. Shirley to A. M. Dole, November 20, 1858; letters to Dole, Ingram, and Kennedy from Bradford and Brother, December 4, 1858, and William T. Bridges, December 29, 1858; to Orrin H. Ingram from D. Kennedy, January 25, 1866, L. B. Howard, November 23, 1869, Knapp, Stout & Co., November 30, 1869, M. Garland, May 22, 1876; and to Ingram and Kennedy from C. H. Eaton, July 11, 1866, all in the Ingram Papers; Ingram, *Letters of a Pioneer,* 7–8; *Chippewa Herald,* December 5, 12, 1873, May 12, 1893, February 26, 1897; *Eau Claire Free Press,* August 22, 1867, March 4, 1886; *Wisconsin Lumberman,* 2:284 (June, 1874).

WHi(X3)44751

John H. Knapp, banker and lumberman who founded the firm that eventually became the Knapp, Stout & Co. Lumber Company, one of the largest lumber companies in Wisconsin.

WHi(X3)18351

Two lumberjacks preparing to cut a felled pine into logs. Photo probably by Charland.

chased a half interest in the Menomonie mill from its owner, David Black, who died before the year was out.

Wilson, still burdened with debt and fearful of his creditors, remained only a silent partner for many years. The firm was known as Black and Knapp until 1850, but Wilson's interest was protected by a secret agreement with Knapp. The death of Black almost immediately after the formation of the new firm had created several problems. Knapp managed to avoid the liquidation that usually followed the death of one member of a partnership by an agreement with the administrators of the Black estate. When Black had sold part of his mill to Knapp, the land upon which it was located had not yet been surveyed. The deed received by Knapp included a stipulation that when the land came on the market, Black would either prove up a pre-emption claim or buy it at the regular land sale and convey half of it to Knapp. Black's death before the survey was made prevented the execution of this agreement. In 1851, after the land had been surveyed, Wilson entered a pre-emption claim on the millsite and immediately conveyed the title to Knapp. In the meantime, in 1850, the administrators of the Black estate had allowed Knapp to buy the entire business. In the years thereafter Knapp sold shares to other partners: Andrew Tainter in 1850, Henry Stout in 1853, and Thomas Wilson, son of William, in 1854. William Wilson's share was carried by Thomas, who signed a declaration of trust. Some years later Knapp admitted his nephew John H. Douglas into the firm. This brought the number of partners to six, including William Wilson. In 1853 the firm name became Knapp, Stout and Company.[5]

When the company was organized in 1846 its mill had a daily capacity of about five thousand board feet, and most of the labor was performed by the owners. From this small beginning the company developed into the largest lumber concern in the country. In 1850 it built a new mill, to which a gang saw was added

[5] Randall, "Chippewa Valley History," in the *Eau Claire Free Press*, July 30, August 6, 1874; Hotchkiss, *Lumber and Forest Industry*, 416, 489, 490; articles by Charles E. Freeman on the history of Knapp, Stout & Co. in the Chetek *Alert*, August 18, October 6, 1916, and in the *Dunn County News*, November 16, 1911, January 10, 1918.

in 1852. When fire destroyed the plant in 1856, the company replaced it with a larger one, which was used until the nineties. In 1862 the owners described themselves as "wholesale and retail dealers in all kinds of merchandise," "manufacturers of and dealers in lumber, lath, shingles, etc.," "flour millers and grain brokers, and dealers in real estate." By the seventies the partners had built up a huge and complex organization having virtually complete control of the Menomonie (Red Cedar) River valley. Its power and flooding dams occupied nearly every available site on the main river and its tributaries. Knapp, Stout and Company logging camps were scattered throughout the north woods. On the routes from the mill to the camps the company operated farms or large gardens to provide food for its workers. At important points down the Mississippi as far as St. Louis it had sales yards and finishing mills. A company steamer returned the rafting crews to their base. In 1870 the regular daily output of its main mill at Menomonie was three hundred thousand board feet of lumber, and a second mill turned out quantities of shingles and laths. The concern operated also a boardinghouse that served two barrels of pork every day, a gristmill and grain warehouse, a general store that did an annual business of three-quarters of a million dollars, a foundry, a machine shop, a wagon and cooperage shop, and a pork-packing establishment. In 1877 its payroll listed fifteen hundred men. It also maintained stores, farms, mills, and warehouses at other places. During the depression following the panic of 1873 this huge and diverse enterprise was able to make a profit when other firms were facing bankruptcy.[6]

The company continued to expand its activities in many directions. In 1878 the partnership became a corporation under the name of "The Knapp, Stout & Co. Lumber Company" with a capital of two million dollars, the ownership remaining the same. Its total yearly output increased from fifty million board

[6] Knapp, Stout & Co. to Dole, Ingram and Kennedy, July 24, 1862, in the Ingram Papers; article by Freeman in the *Dunn County News,* January 10, 1918; *Chippewa Falls Democrat,* March 17, 1870; *Wisconsin Lumberman,* 1:20, 205–206 (October, 1873; January, 1874); *Northwestern Lumberman,* February 17, 1877; Hotchkiss, *Lumber and Forest Industry,* 489.

feet of lumber in 1871 to double that amount in 1879. This was, according to a contemporary trade journal, probably more than any two other firms in the United States were capable of producing; yet only three years later it had an output of a hundred and fifty million board feet. The company invested all its surplus capital in land, buying heavily from Cornell University and the railroads of northern Wisconsin. It promoted urban communities on these lands, which proved to be highly profitable. Rice Lake, the seat of Barron County, was founded and developed by the Knapp-Stout firm. In 1879 it was planning to build a sawmill at St. Louis and raft logs down the Mississippi with its own steam tugs. At the same time it was using the railroads extensively for shipping finished and seasoned lumber to the valley of the Missouri. During the eighties it swallowed up several smaller concerns. In 1882 the authorized capital was increased to four million dollars. By 1890, foreseeing the disappearance of white pine in Wisconsin, the company had begun to buy pinelands in Missouri, Arkansas, and Mississippi. In 1891 there were rumors that it would move into the South, and an Eastern syndicate actually acquired an option to buy the property at a price of nearly eight million dollars, but the purchase was not consummated. In 1898 the daily sawing capacity of its several mills was three-quarters of a million feet. For fifteen years before 1896 it had an average of two thousand men on its payroll. Its production during the half century 1846–96 totaled nearly two billion feet of lumber. As a firm it paid a million and a half dollars in state taxes over the thirty-five-year period ending in 1896.[7]

Knapp, Stout and Company represented what was perhaps the maximum growth of which a single concern was capable. It came as near being a monopolistic unit as the economics of the

[7] *La Crosse Leader*, January 14, 1871; *Chippewa Herald*, June 4, 1875, May 3, 1878, March 24, 1882, February 16, 1883, May 8, June 5, 1891; *Eau Claire Free Press*, June 25, 1885, May 7, 14, 1891, February 15, 1894, May 18, 1899; article by Freeman in the *Dunn County News*, January 10, 1918; *Wisconsin Lumberman*, 1:373 (March, 1874); 2:259, 364 (June, July, 1874); *Northwestern Lumberman*, March 22, June 7, 1879; summary of the history of Knapp, Stout & Co. in the "Question Box" section of the *Wisconsin Magazine of History*, 3:469–471 (June, 1920); Hotchkiss, *Lumber and Forest Industry*, 489–490. Cornell University had acquired timberland in Wisconsin in exchange for Morrill Act scrip. See below, pages 171–172.

lumber industry permitted. The development of companies that would control all the manufacturing or marketing activities of a region had been dreamed of by Wisconsin lumbermen at least as early as 1850. In 1851 Moses M. Strong had plans for establishing a company that would own every millsite on the Wisconsin River "so as to give the company a complete monopoly of the whole pine timber business on that river." In 1874 a Minneapolis syndicate attempted to corner all the lumber on the upper Mississippi. In the mid-eighties it was claimed by "outside" manufacturers that Oshkosh dealers controlled most of the lumber in that "section of the country." And in 1890 it was rumored that the lumbermen of Eau Claire were planning to obtain control of the entire Middle Western market. But none of these things was possible in an industry which normally obtained the highest rate of profit from a mill of comparatively small size. Even Knapp-Stout could not concentrate its entire production in a single mill.[8]

The expansion of the individual lumbering company was, however, accompanied by another development in the direction of monopolistic control, the organization of numerous firms into associations for the purpose of influencing prices. No serious attempt had been made to form any such combination to overcome the slump in prices that followed the panic of 1857, but early in 1874, when prices were again low, markets glutted, and bankruptcies numerous, there began to be talk of creating an organization to "control the market efficiently and intelligently." The crucial problem was the disparity between supply and demand: consumers were unable to purchase all the lumber produced annually. To the lumbermen the solution seemed simple; if the supply on the market was reduced until it reached parity with the demand, the price would automatically be restored to its normal level. Accordingly the main purpose of the first lumbermen's associations was to reduce output.

[8] Memorandum about Speculations at East Relative to Saw Mills, 1851, in the Strong Lumber Papers; La Crosse *Republican and Leader,* March 8, 1874; Oshkosh *Weekly Northwestern,* October 7, 1886; *Eau Claire Free Press,* November 20, 27, 1890.

In July, 1874, several Wisconsin lumbermen attended a preliminary convention at Williamsport, Pennsylvania, which laid the groundwork for the formation of the National Association of Lumbermen. Membership in the association was limited to such persons "as are engaged in the manufacture of lumber or timber of any kind, round, hewn or sawed, or who may be the owner of timber lands." At the same time Canadian lumbermen met at Ottawa. Both groups expressed the view that lumber production should be cut at least fifty per cent the following year.

The American association planned to reconvene at Saginaw, Michigan, on September 16. Meanwhile local associations were formed in various districts of Wisconsin to cooperate with the national association and to select delegates to the Saginaw meeting. Conservation of natural resources became the avowed purpose of the lumbermen, who hoped to win public approval for their inevitably unpopular price-fixing activities by making them secondary to curbing the "present senseless slaughter of pine." Some groups did admit openly that they sought to eliminate competition, set prices, and secure a monopoly on timberlands, but only as a means of curbing wasteful destruction of the nation's forest resources. The variety of benefits that certain lumbermen hoped to gain from a central organization was revealed by the pledges which the Chippewa valley delegation made before they left for the Saginaw meeting: they promised to obtain a limitation of the next season's output, reform in the taxation of timberlands, regulation of lumber inspection, and regulation of lumber measurements.

When the convention assembled, the most immediate threat to the success of any curtailment plan was President Grant's new reciprocity treaty, which would admit Canadian timber products into the United States duty-free. After much debate the association adopted resolutions condemning the proposed treaty and recommending that the next season's output be reduced at least a third below that of the preceding year. Joseph G. Thorp, an Eau Claire lumberman, became the first president of the association and Thaddeus C. Pound of Chippewa Falls one of the vice-presidents. The other officers, too, reflected the dominant position of the Middle West in the organization.

The curtailment plan failed to meet expectations. Although firm after firm announced during October that it would reduce the amount of its cut for the winter, the association was forced to issue a circular at the close of the month stating that the sentiment of the country was against curtailment and that, in accordance with a strict interpretation of the resolution adopted at Saginaw, those who had agreed to restrict their output were no longer bound to do so. Many companies continued to reduce production, but by December it had become clear that the plan had failed. The market was again greatly overstocked. Chicago had seventy-five million board feet more than it had held twelve months before.

Lack of snow during the winter of 1874–75 forced some curtailment in that it prevented loggers from putting their usual quota of logs into the driving streams. The lumbermen felt that providence had come to their aid, and enthusiastically began a new campaign for restriction of production. When the association met in Chicago in September, Thorp declared in vigorous, if somewhat high-flown language that the time had come to stop the "sacrificing of the 'whispering pines and hemlocks' with the ruthless hand of the American vandal in his search for wealth."

But rhetoric was no cure for the weaknesses of the faltering association. The lumbermen discovered, as had the railroads and other industries, that voluntary price-fixing agreements could not be depended upon to control the market. It was too much to expect that profit-minded lumbermen, fully aware that their fortunes rested upon continued inroads into the forest, would be inspired with a fervor for curtailment by idealistic and possibly hypocritical expressions of conservation sentiment. Concern for his bank account could make a practical conservationist of a hardheaded lumberman, but he was not prepared to trust an organization to decide for him what his output should be. In 1876 the National Association again issued a statement that there was need for curtailment of production, but it was futile. The association had failed to strike the right note and to get members from different sections to abide by an agreement on policy. It was soon to be replaced by new and more vigorous organizations.[9]

The local organizations that emerged from this national movement for a curtailment program proved to be of more immediate service to the lumbermen than the larger association. Typical among these was the Chippewa Valley Lumbermen's Exchange, established in 1875 by the principal manufacturers and loggers of Eau Claire, Chippewa Falls, and Menomonie. This organization not only cooperated wholeheartedly with the national one, but also planned and carried out a local program. Its main objective was to eliminate disastrous competition among its members by getting general agreement on price lists. To prevent friction among them it offered to arbitrate disputes and thus forestall litigation. A good organization could also exert pressure on local government to refrain from imposing high taxes and indulging in "harassing expenditures." When occasion arose, the Exchange acted as a medium of communication between lumbermen of the Chippewa valley and those of other sections. Its efforts along these lines were successful enough to maintain its existence long after the national organization had disappeared.[10]

But it soon became apparent that some kind of organization on a large scale had to be developed if the market was to be controlled. It was demonstrated in 1879 when the lumbermen of the Oshkosh district made an effort to raise the price of lumber a dollar, and again the next year when the Chippewa valley lumbermen tried to cooperate with Chicago dealers to bring about "an immediate and positive advance in prices." Recognizing the need for broader action, the manufacturers of Michigan,

[9] La Crosse *Republican and Leader,* February 28, September 12, 26, 1874; *Eau Claire Free Press,* July 9, September 24, 1874, January 21, September 23, 1875; *Chippewa Herald,* August 7, 28, September 11, 18, 25, October 16, 23, November 20, December 4, 11, 1874, August 20, 1875, August 25, September 8, 29, 1876; circular of the National Association of Lumbermen, October 26, 1874, and G. E. Stockbridge to Orrin H. Ingram, [October, 1875], in the Ingram Papers.

[10] *Eau Claire Free Press,* May 13, September 2, 1875, March 16, August 17, 1876, March 2, 1882, February 1, 1883, May 28, 1885, March 3, 1887; *Chippewa Herald,* May 14, September 3, 1875.

Wisconsin, and the Mississippi Valley in 1882 organized the Lumber Manufacturers' Association of the Northwest.[11]

Based on a regional foundation, the new association had a better chance of succeeding than the old national one had had. Its members faced the same problems and, theoretically at least, should be able to agree on policy. In its first effort to curtail production, however, the association encountered the same fundamental difficulty that had disabled the older one. In 1884 the officers of the association decided that members must shut down their mills by October 15, but that it would be well to withhold announcement of that decision until September 1 in order to insure that millowners would not work double time to build up a surplus stock to sell after the mills had closed. Rumors of the plan leaked out in advance, but when the association met in convention in August its members voted to shut down their mills on or before November 1 and keep them closed for a month. They recognized that to make this plan effective they must have the cooperation of the independent loggers, but they saw no way of persuading them to remain idle. Nor did they themselves abide by the agreement. During the month of October many mills operated day and night, each owner hoping thus to obtain an advantage over those of his competitors who might shut down on the date agreed upon.

In January, 1885, the secretary of the association issued a circular urging that logging operations be restricted to forestall an overstock in the spring, but his advice seems to have gone unheeded. When the association met late in May it again voted to reduce production, but permitted each member to decide how and when he would do so, thus tacitly admitting the practical failure of this method of handling the problem. The comment of a delegate to the convention is revealing: "They didn't agree on anything, and it seems to me they didn't want to. Oh, yes, they want to curtail, but not a man of 'em would pledge himself to any method that would mean curtailment. That was the it of it, and I can tell you why. Men who took a prominent

[11] Oshkosh *Northwestern,* August 21, 1879; *Eau Claire Free Press,* February 12, 1880; Hotchkiss, *Lumber and Forest Industry,* 309.

part in the meeting don't mean to curtail—won't curtail just as long as they can sell *their* lumber. You couldn't bind them to any agreement with a log chain. For the good of the white pine trade at large they don't care the snap of their finger." He had obviously sized up the situation correctly. By July few millowners had made any move to cut their production schedules. At another meeting in September they again voted for curtailment, but again without effect. By 1886 the problem had solved itself, at least for the moment. General prosperity had returned to the country, and the president of the association announced that there was no overproduction in the lumber industry.[12]

The largest and perhaps most successful regional organization was the Mississippi Valley Lumbermen's Association, which was founded in September, 1891. By 1905 its members represented ninety per cent of all the white pine production in Wisconsin and Minnesota and areas tributary to the Mississippi; at one time they manufactured more than half the white pine lumber marketed in the United States. Having learned from the failure of earlier organizations that it was impossible to control the market simply by attempting to curtail production, the new organization set about fixing prices directly. At its first meeting in 1891 it decided to advance the price of lumber in the territory it controlled. Lumbermen who would almost certainly have violated a curtailment agreement found it comparatively easy to adhere to a price list previously agreed upon. Their compliance was facilitated by the fact that large syndicates and combines had already been formed by some companies in the Mississippi Valley.[13]

So successful was the Mississippi Valley Lumbermen's Association in raising prices that by January of 1892 a federal grand jury at Minneapolis was investigating the organization to determine whether it was a combination in restraint of trade, as defined by the recently enacted Sherman Antitrust Act. In Febru-

[12] A. G. Van Schaick to Orrin H. Ingram, June 30, 1884, in the Ingram Papers; *Chippewa Herald,* February 22, July 11, August 29, 1884; *Eau Claire Free Press,* August 7, September 4, October 16, 1884, January 22, May 28, June 4, 11, July 30, September 24, 1885, September 30, 1886.

[13] *Eau Claire Free Press,* September 10, 1891; *American Lumbermen,* 1:292–293.

ary the grand jury indicted the principal officers and directors of the association for conspiracy to set prices. The indicted lumbermen, assuming that the law was violated only when an outright monopoly was formed and an unreasonable price agreed upon, were not greatly perturbed. Their defense was that the Sherman law was not intended to apply to lumbermen who met "in good faith for the purpose of getting at the condition of the lumber market." Their unruffled attitude was not without justification. The district court refused to find them guilty, and in August the association again raised the price of lumber.[14]

The association grew and prospered. In the season of 1892–93 its members included the manufacturers of more than four billion board feet of white pine lumber. It was not always successful in maintaining the price advances that had been agreed upon, but it was able to strengthen its position by inducing smaller organizations and individual manufacturers to abide by its prices even though they did not became members. Ordinarily a price list was not binding unless eighty-five per cent of the membership had agreed upon it, but occasionally the officers announced as binding a list that had been accepted by only eighty per cent. In 1895 the association had the support of seventy members, thirteen non-members sympathetic with the purposes of the organization, and twenty-five members of the Wisconsin Valley Lumbermen's Association that accepted the prices agreed upon by the larger group.[15]

The Wisconsin Valley Lumbermen's Association, founded in 1895, was one of the district associations that was organized to do on a smaller scale what the Mississippi Valley Association was doing for the Middle West as a whole. Organized for the imme-

[14] United States *v.* Nelson *et al.*, 52 *Federal Reporter* 646 (1892); *Eau Claire Free Press,* January 21, 28, February 18, March 24, August 18, 1892. "An agreement between a number of dealers and manufacturers to raise prices, unless they practically control the entire commodity, cannot operate as a restraint upon trade, nor does it tend to injuriously affect the public," wrote the federal district judge in the Nelson case. 52 *Federal Reporter* 647.

[15] *Chippewa Herald,* August 18, 1893, March 2, 1900; *Eau Claire Free Press,* August 29, 1895, August 30, 1900; Circular Letter of A. L. Ulrich, Chairman of the Price-List Committee of the Mississippi Valley Lumbermen's Association, September 12, 1895, in the Ingram Papers; Bureau of Corporations, *The Lumber Industry,* pt. 4, p. 494.

diate purpose of establishing and maintaining a grading system, it also participated in price-fixing activities, some of which have been described. In 1906 it amalgamated with the Mississippi Valley Lumbermen's Association to form the Northern Pine Manufacturers' Association. Among the other district associations was the Lake Superior Lumber Manufacturers' Association, established in 1900 by the twenty-six leading sawmill companies along the lake.

Nor were the price-fixing organizations limited to sawmill owners. In 1880 the sash, door, and blind makers organized the Northern Wisconsin Manufacturing Association, which circulated uniform price lists and sought to prevent overproduction. In 1894 some twenty manufacturers signed the articles of agreement governing the Northwestern Stave and Heading Association, another price-fixing agency. Retail dealers combined in the Wisconsin Lumber Dealers' Association, whose function was to evaluate dealers and keep wholesalers and manufacturers from selling to those it did not find acceptable. Membership in such organizations enabled dealers in large distributing centers like Chicago and Minneapolis to exert considerable influence on market conditions. Their organizations, the manufacturers insisted, were not interfering with the free course of trade. "We are not," said a prominent member of the Northwestern Stave and Heading Association in 1894, "getting up a trust or anything of that sort. We simply wish to establish uniformity of prices."[16]

Two methods of fixing uniform prices were developed before 1910. Until 1906 most associations merely circulated price lists, to which their members were expected to adhere. About that time federal and state governments, influenced by President Roosevelt's "trust-busting" philosophy, began a series of prosecutions that worried the lumbermen. To meet this threat the associa-

[16] *Eau Claire Free Press,* February 2, March 9, 1882, February 15, 1883, May 21, July 23, 1889, February 15, 1894, July 27, 1899, June 7, 1900; *Chippewa Herald,* May 29, 1891, May 12, 1893; Oshkosh *Northwestern,* September 30, 1880; Wisconsin Lumber Dealers' Association, *Constitution and By-Laws, Lists of Active and Honorary Members, Minutes and General Information* (Elkhorn, Wisconsin, 1891); B. F. Nelson to Empire Lumber Co., April 20, 1885, in the Ingram Papers.

tions issued their price lists under the guise of "market reports" or operated as open-price associations. Under the latter system each mill could set its own price, but it received accurate information on the state of the market and on costs of manufacture from the organization to which it belonged.[17]

Price lists were sometimes very effective methods of price-fixing. Many factors, on the other hand, tended to defeat the price schedules of the associations. Members who cut their prices were by no means uncommon. Lumbermen in debt were usually forced to sell their product for whatever it would bring. The difficulty of grading lumber accurately, the physical isolation of the sawmills, and the competition between different kinds of wood all complicated the problem. The steady rise in lumber prices after 1900 was largely the result of factors other than organized price-fixing: depreciation of the currency, exhaustion of readily accessible supplies of timber, and increasing transportation costs.[18]

Curtailment of production, usually a necessary adjunct to price-raising, was always more difficult to accomplish than the circulation of price lists. After the federal government began its antitrust activity, associations desiring to reduce the amount of timber on the market had to rely on propaganda disseminated by the trade papers, since it was no longer possible to adopt resolutions at meetings or circulate agreements among members. Frequently, too, curtailment campaigns were carried on by association members "acting as individuals." But it was always difficult to get the cooperation necessary to the success of the program. Reduction of output did not impose too serious a sacrifice on the debt-free operator who had only his own interests to consider, nor on operators owning timberland, who could count on the natural increase in its value to offset a loss elsewhere; but most mill operators were in no position to close their establishments at will. Some of the curtailment campaigns were successful, but they effected less reduction of output than

[17] Ise, *United States Forest Policy,* 338–340.

[18] Oshkosh *Weekly Northwestern,* January 8, 1891; Ise, *United States Forest Policy,* 343–348.

did the inevitable shutdowns accompanying recurrent periods of depression.[19]

Whether the lumbermen's organizations actually created monopolistic conditions is an open question. Some of them certainly did raise prices temporarily above the level justified by market conditions, but they were prevented from enjoying consistent success by the difficulties inherent in the voluntary association. Wisconsin in particular never had the tight organizations that some other regions developed, with the possible exceptions of the Mississippi Valley Lumbermen's Association and the Northern Hemlock and Hardwood Manufacturers' Association.

The government's moves to put an end to the monopolistic activities of lumbermen's associations failed except as they forced the associations to be more circumspect. In 1893 the federal district court at St. Paul declined to vacate a temporary injunction forbidding the Northwestern Lumbermen's Association to advise dealers not to buy of one of its members who had allegedly violated the rules of the organization, but the association continued to issue its price lists unrestrained. In 1907 the Wisconsin legislature appointed a joint committee to investigate the lumber industry, especially "combinations, pools and agreements intended to restrain or prevent competition in the business of buying, selling and dealing in lumber or to fix or regulate the price thereof." For reasons unexplained the report of this committee, if it made one, never appeared in the usual places of publication. In 1932 a legislative committee praised the activities of lumbering associations in fostering cooperation and "combatting competition."

Lumbermen always insisted that the organization of their industry was inimical to monopoly. The National Lumber Convention of 1897 reported that "in no branch of manufacture is competition fiercer. Any combination to restrict production or control prices is absolutely impossible by reason of the fact that there are over *twenty-three thousand lumber manufacturing in-*

[19] Federal Trade Commission, *Conference with National Lumber Manufacturers' Association* (Chicago, 1915), 138; Ise, *United States Forest Policy*, 350–353.

stitutions in the United States, and located in every State and Territory." The lumberman could defend the existence of his associations on grounds other than control of the market. They advertised lumber products, established systems of grading and inspection, maintained credit bureaus, investigated freight rates and traffic problems, maintained employment bureaus and studied labor conditions, sought to improve the quality of products, encouraged better methods of accounting, advised the public about conditions in the industry, studied forest management, and agitated for better methods of taxing lands.[20]

That they were capable of maintaining a rather tight control of prices at times is revealed in the history of the Northern Hemlock and Hardwood Manufacturers' Association, founded in 1910 by a merger of the Northwestern Hemlock Manufacturers' Association and the Hardwood Manufacturers' Association of Wisconsin. By 1920 its membership included some seventy-five firms, largely in Wisconsin and upper Michigan. It was being conducted, according to the Federal Trade Commission, "largely as a medium to facilitate agreements and understandings among themselves for the substantial lessening and frequently the practical destruction of price competition"; the leaders of the association had themselves "paid strong and frequent tribute to their organization as a powerful aid and factor in maintaining prices on a weak market and advancing them when conditions were favorable."[21]

The inherent weakness of voluntary combinations was always the difficulty of forcing compliance upon half-hearted members. In the Wisconsin lumber industry two forms of organization existed that avoided this problem. The first of these was the interlocking management. Orrin H. Ingram of Eau Claire, for instance, was at one time a director of the Chippewa Logging Company, vice-president of the Chippewa Lumber and Boom

[20] *Session Laws,* 1907, p. 1286; *Eau Claire Free Press,* January 5, 1893; *Chippewa Herald,* January 11, 1900; [Wisconsin] Committee on Land Use and Forestry, *Forest Land Use in Wisconsin* (Madison, 1932), 46–47; *Memorial of the National Lumber Convention to the Senate Finance Committee,* 2–3; Ise, *United States Forest Policy,* 334–335.

[21] *Report of the Federal Trade Commission on Northern Hemlock and Hardwood Manufacturers' Association* (Washington, 1923), viii, x–xii.

WHi(X3)37690

A crew of sawyers with their tools, about 1900. Notice the sledge-hammers for driving wedges and the eight-foot sticks for measuring standard lengths of logs.

WHi(X3)145

Orrin H. Ingram, a prominent Wisconsin financier and lumberman who also had extensive interests in lumbering outside Wisconsin (see Index under "Ingram").

Company, president of the Rice Lake Lumber Company, president of the Empire Lumber Company of Eau Claire, principal stockholder in the Empire Lumber Company of Minnesota, vice-president of the Standard Lumber Company of Dubuque, and a heavy stockholder in the Weyerhaeuser Timber Company, a holding company for Pacific coast timberlands. He also invested privately in Midwestern, Southern, and Pacific coast timberlands. He was likewise president of the Eau Claire National Bank and the Eau Claire Water Works Company, director and treasurer of the Canadian Anthracite Coal Company, and an "officer in many other large enterprises." Frederick Weyerhaeuser at one time had a voice in the management of eighteen manufacturing concerns in Minnesota, Wisconsin, and Illinois. More important, he was the controlling spirit of the Mississippi River Logging Company and its affiliate corporation, the Beef Slough Manufacturing, Booming, Log Driving and Transportation Company. Together these two companies represent the second type of combination. They were fundamentally syndicates composed of the most important lumbering companies of the Chippewa and Mississippi River manufacturing areas. They consolidated all the timberland and logging interests of the large mills below Lake Pepin and controlled almost the entire log output of the Chippewa River.[22]

The Mississippi River Logging Company was potentially the most monopolistic enterprise developed by lumbermen operating in Wisconsin. Its great power stemmed from its policy of getting control of timberlands as well as manufacturing units. The acquisition of concentrated timber holdings was of greater significance in the development of monopolistic enterprise than combinations among lumbermen or large manufacturing plants. Voluntary lumbermen's associations were not strong enough to monopolize production and distribution in any long-term program. Manufacturing plants were limited in size by the economics of the industry; they could effect the usual economies of large-scale production but were unable to take full advantage

[22] *American Lumbermen,* 1:321; Hotchkiss, *Lumber and Forest Industry,* 606. On Weyerhaeuser see also pages 146–160 below.

of them because of the relatively high cost of transportation. A large mill soon used up the nearby timber, and the cost of transporting raw material from a distance soon offset the advantages achieved by large-scale production. Moreover, a sawmill which had not yet paid for itself could not be moved except at the sacrifice of much of the original investment. And the larger the mill the more likely it was to exhaust its immediate supply of timber and incur higher transportation expenses before it had paid for itself. A company that owned most of the timber in its area, on the other hand, not only made a good profit from the increase in land values, but could dictate to all manufacturing concerns dependent upon it for logs. The Mississippi River Logging Company was organized primarily to buy land for its constituent firms, but because they had common interests it could also serve to some extent as a price-fixing agency. The size and power of this great logging syndicate made it the most significant of all the lumbermen's organizations operating in Wisconsin.[23]

[23] Bureau of Corporations, *The Lumber Industry,* pt. 1, pp. 3–4.

9. The Mississippi River Logging Company

THE natural correlate of log rafting on the Mississippi was the growth of sawmilling centers drawing their raw materials from the streams of Wisconsin and Minnesota. The first mills on the river were built as early as the 1850's, when the St. Croix and Black River areas were the chief sources of timber. These lumber-manufacturing centers at a great distance from the forests created two conflicting interests in the pineries of Wisconsin: the log drivers and landowners who supplied the raw materials to the distant mills, and the local manufacturers who regarded it as their prerogative to process all the logs cut in their district.

The log drivers argued that their operations enhanced the value of the lands which furnished them their logs and likewise gave employment to local laborers and farmers. They viewed their opponents as "overgrown monopolists who have become rich by trespassing on Government lands, and otherwise enjoy privileges denied to men of limited means." The activities of an outside interest would not drive the local mills out of business but would merely furnish healthy competition that would enable logging contractors to sell their logs at a higher price.[1]

[1] Norton, *Mississippi River Logging Company,* 10; *Chippewa Union and Times,* April 27, September 21, 1867.

The local manufacturing interests countered with the charge that not they but the downriver mills were the monopolists, "possessing the capital to swallow up" every small millowner. History had shown, they said, that every country that had permitted its raw materials to be exported for manufacture failed to measure up to other nations in "wealth, power, influence, progress in civilization, and refinement," and that since the support of institutions necessary to political and social progress depended on revenue from all the resources of a state, any policy that denied a community the "advantages to be derived from the manufacture of its raw products or takes them to a distant or foreign shore for that purpose can only be regarded as a grievous blunder committed by its rulers."[2]

This conflict between local and foreign interests within the lumber industry took place not only in the Mississippi River driving area but wherever logs could be sent to distant mills for manufacture into lumber. The Green Bay and Lake Winnebago districts had trouble of this kind, but the clash of greatest magnitude occurred in the Chippewa valley between 1865 and 1880.

About 1865 the lumbermen of the Eau Claire district learned to their great alarm that speculators from Michigan and the Wisconsin cities of Oshkosh and Fond du Lac had begun to buy timberlands in the Chippewa valley with the obvious intention of shipping great quantities of logs downriver to mills on the Mississippi. As a countermeasure a group of local millowners in 1866 purchased the land around the entrance of Beef Slough, one of the mouths of the Chippewa and the only feasible natural harbor for the sorting, storage, and rafting of logs destined for Mississippi River points. They obtained from the Wisconsin legislature a charter for a corporation having the exclusive right to construct improvements there. Their purpose, of course, was not to build works themselves but to keep others from doing so.[3]

Fully aware of the disastrous implications for them, the outside capitalists rejected the invitation of the local interests to

[2] *Eau Claire Free Press,* September 5, 1867. The issue for January 28, 1875, contains a historical sketch of the Beef Slough enterprise.

[3] *Eau Claire Free Press,* August 1, 1867, January 28, 1875; *Chippewa Herald,* August 14, 1874; Hotchkiss, *Lumber and Forest Industry,* 477; *Private and Local*

invest in Chippewa valley sawmills and, with the promise of support from the Mississippi River millowners, resolved to fight. Under the general incorporation law of Wisconsin the speculators organized in 1867 the Beef Slough Manufacturing, Booming, Log Driving and Transportation Company. The stated purpose of the new corporation—to build a sheer boom at the head of the slough, a jam boom near its mouth, and suitable works for sorting and handling logs—clearly violated the exclusive rights previously granted to the Chippewa valley group. Nevertheless the impatient outsiders, eager for some profit on their investment, immediately sent an agent to start building works at the slough. The necessary legislative authorization could be obtained later.[4]

Both business groups now marshaled their lobbyists and sympathizers at the state capitol. Amid unusual excitement the Assembly passed a bill to grant special privileges to the new Beef Slough Company, but it was rejected by the Senate. The exuberant Chippewa lumbermen celebrated their precarious victory by forcibly halting the work of the new company and building a dam of their own across the head of the slough, ostensibly to improve navigation but in reality effectively sealing it off. After applying for an injunction against the millowners for obstructing a navigable stream, the Beef Slough Company removed the offending dam before their suit could be heard in court.[5]

In February, 1868, the rival logging interests returned to the struggle in the legislature. Heavily freighted with political significance, the Beef Slough charter bill failed this time to pass

Laws, 1866, ch. 181. An earlier charter, granted in 1855 to the Chippewa River Boom Company, a concern controlled by local interests, might have been construed to give the company exclusive control of the slough, but the Chippewa millowners wanted to be doubly sure. *Private and Local Laws,* 1855, ch. 331. Beef Slough left the main channel of the Chippewa about three miles below Durand and entered the Mississippi about twelve miles below the true mouth of the Chippewa. Norton, *Mississippi River Logging Company,* 11.

[4] *Eau Claire Free Press,* January 28, 1875; *Circuit Court of the United States for the Western District of Wisconsin . . . in Equity: Information and Bill of Complaint in the Case of U. S. v. Beef Slough Manufacturing, Booming, Log Driving and Transportation Company . . .* (n.p., 1878), not paginated; Hotchkiss, *Lumber and Forest Industry,* 481.

[5] *Assembly Journal,* 1867, pp. 160, 450, 694, 1001; *Senate Journal,* 1867, pp. 587, 790, 842; *Chippewa Union and Times,* January 26, March 16, 23, April 6, 27,

even the Assembly, though only a single member of the Committee on Lumber and Manufactures, to which it had been referred, reported against it. But the exultation of the Eau Claire lumbermen was short-lived. While they were holding a mass meeting to celebrate the death of the bill, the Assembly was considering another "to incorporate the Portage City Gas Light Company," which, with virtually no opposition, became law on March 24. Buried in the new act was an apparently innocuous rider stipulating that "where any franchise or privilege has been or shall be granted by law to several persons, the grant shall be deemed several as well as joint, so that one or more may accept and exercise the franchise as though the same were granted to him or them alone." Although it still lacked a definitive charter, the logging company had won its fight. Among the persons named in the charter granted to the local millowners in 1866 was one James H. Bacon, who had since shifted his allegiance to the logging company and who now assumed the right to build improvements in Beef Slough as if it had been granted to him alone. This right he immediately transferred to his associates in the logging company.[6]

The local millowners were not yet disposed to give up the struggle. In the spring drive of 1868 the entire fifty million board feet of logs for which the Beef Slough Company had contracted came to rest in the booms of the Eau Claire sawmills, where they were detained on the ground that the local sorting works were unable to separate the logs destined for the slough from those belonging to Eau Claire. Ordinarily in a situation of this kind the parties agreed upon an exchange of logs of equal value, but now each side claimed that the other flatly refused to make an exchange.[7]

May 11, July 27, September 21, 1867, January 18, February 1, 1868; *Eau Claire Free Press,* May 16, August 1, 22, September 19, 1867, January 30, 1868, January 28, 1875; Merk, *Economic History of Wisconsin during the Civil War Decade,* 91–92.

[6] *Assembly Journal,* 1868, pp. 314, 608, 630, 694; *Eau Claire Free Press,* January 16, 30, February 20, March 5, 12, 1868; *Chippewa Union and Times,* March 14, 1868; *Private and Local Laws,* 1868, ch. 331; Merk, *Economic History of Wisconsin during the Civil War Decade,* 93.

[7] *Chippewa Union and Times,* April 25, May 2, 1868; *Eau Claire Free Press,* May 7, 1868, January 28, 1875.

Exasperated by the delay that threatened it with financial ruin, the logging company resorted to boom cutting, a procedure that permitted the logs belonging to the local mills to move down the river with its own. On one occasion nearly twelve million board feet of logs were released from a single boom in Eau Claire County. These tactics ended when a sheriff's posse composed largely of Eau Claire "property owners" arrested Bacon and the crew of a hundred and fifty drivers who were cutting the booms. Bacon and his chief lieutenant were bound over in bonds of ten thousand dollars each to keep the peace for six months. Eventually an uneasy truce was worked out which enabled Bacon to finish the drive, but as late as June many of the Beef Slough logs remained in the Eau Claire booms. Unpleasant bickering occupied both sides throughout the next two years.[8]

Finally, in March of 1870, the legislature explicitly authorized the logging company to maintain the booms and piers it had thus far constructed along the Chippewa and in Beef Slough, and granted to it the exclusive privilege of driving and booming all logs, timber, ties, and fenceposts in Beef Slough, delivering them to their owners, and charging fees for these services. But the charter came as an anticlimax, for the opposition of the Chippewa River lumbermen had already bankrupted the concern. The financial backers of the company and the lumber companies dependent upon its services now began to look for a new group of capitalists to take over the enterprise.[9]

They had not long to wait. Certain lumbermen of Rock Island and Davenport who had been operating on the Black and St. Croix rivers began to explore the possibility of developing

[8] James Ginsell to Ingram and Kennedy, June 17, 1868, Orrin H. Ingram to Kennedy, August 25, 1868, and Cousins and Bartlett to Ingram, February 27, 1869, in the Ingram Papers; *Eau Claire Free Press,* May 7, July 2, 9, 1868, February 11, May 6, 1869; *Chippewa Union and Times,* May 9, 16, June 13, 1868. In 1869 the millowners managed to get a law levying a heavy fine for placing refuse lumber or other materials in any "meandered navigable stream" if such an act might be "calculated to impede the free navigation thereof." *General Laws,* 1869, ch. 152, sec. 112. This law, they believed, could be used against the logging company.

[9] *Private and Local Laws,* 1870, ch. 299; *Eau Claire Free Press,* January 28. 1875; Norton, *Mississippi River Logging Company,* 12.

the Chippewa. The driving spirit behind this movement was Frederick Weyerhaeuser, destined to become the greatest lumber magnate of the century. A native of southern Germany, Weyerhaeuser had come to the United States in 1852 at the age of seventeen. After spending four years in Erie County, Pennsylvania, he moved to Rock Island, where he obtained employment in a grain and lumber business. He soon formed a partnership with his brother-in-law, Frederick C. A. Denckmann, and bought a part ownership in a sawmill at Rock Island. So well did this partnership prosper that by 1870 the company was running two sizeable mills, and Weyerhaeuser had become highly esteemed for his business acumen by the lumbering fraternity up and down the river.[10]

About 1868 Weyerhaeuser started to buy logs experimentally in the Chippewa district. These transactions proved to be exceptionally profitable, but like the others who were driving logs to distant mills he suffered from the obstructive tactics of the Chippewa millowners. When the Beef Slough Company reached the point of dissolution Weyerhaeuser was concerned; he and the other Mississippi River millowners who used Chippewa logs could not dispense with its services. Realizing also that the successful prosecution of the booming business at the slough would yield high profits and place its managers in a position of power, he induced Lorenzo Schricker and Elijah Swift, Mississippi River lumbermen, to join him in 1870 in leasing the property of the Beef Slough Company for a two-year term. Before the end of the year nearly all the substantial milling firms along the Mississippi between Beef Slough and St. Louis had been drawn into the enterprise.

During the same year several meetings of the firms operating at Beef Slough laid the groundwork for the establishment of a permanent organization, and in December the Mississippi River Logging Company was created at a gathering of lumbermen at the Briggs House in Chicago. The new company, reputed to represent three-fourths of the sawmilling investment between

[10] Hotchkiss, *Lumber and Forest Industry,* 607; Norton, *Mississippi River Logging Company,* 11–13.

WHi(X32)6320

Fulton, Illinois, 14 July 1901. Log raft on the Mississippi river being maneuvered by the river tugboats *Lydia Van Sant* (near end) and *J.W. Van Sant* (far end), en route from Stillwater, Minnesota, to Muscatine, Iowa. View looking upstream, probably from the Fulton bridge. Note the dory on the end of the log raft for scale.

Photo courtesy Weyerhaeuser Archives

Frederick Weyerhaeuser (1834–1914), a businessman prominent in the history of lumbering and founder of the Weyerhaeuser Company. Photo taken ca. 1900 when Weyerhaeuser was sixty-five years old.

Winona and St. Louis, obtained in Iowa a charter which authorized capitalization at a million dollars. Schricker was elected its president, and as its agent he went immediately to Chippewa Falls to buy logs for the drive of 1871.[11]

The company put more than thirty million board feet of logs into the river during its first year of business, but two-thirds of them were prevented from reaching the slough by continued obstruction and outright violence on the part of the local interests. It soon became apparent, however, that violence could not be employed indefinitely against a syndicate rich enough to put an increasing number of logs into the water each year. Eventually a substantial portion of them must pass the Eau Claire booms for lack of facilities to hold them. The Chippewa valley lumbermen, therefore, had to seek a new weapon. They began to explore the possibilities of the ancient legal doctrine of free navigation. The prospect of fighting the syndicate with lawbooks was inviting, for the basic common law continued to follow the simple dictum of William Blackstone that any navigable stream is a public highway, although in the United States a confusion of jurisdiction had resulted from the superimposition of a structure of organic and statute law and complicating political rivalries.[12]

The doctrine of free navigation had first been extended to the Mississippi Valley by the Ordinance of 1787, which stipulated that "the navigable waters leading into the Mississippi and Saint

[11] Laird, Norton & Co. to Ingram and Kennedy, February 17, June 30, July 6, September 8, 1869, in the Ingram Papers; *Chippewa Falls Democrat,* July 22, 1869, February 2, 1871; *Chippewa Herald,* February 25, 1871; Norton, *Mississippi River Logging Company,* 14, 34. The firms represented at the Chicago meeting were C. Lamb and Sons and W. J. Young & Co. of Clinton, Iowa; Schricker and Mueller and L. S. Davis of Davenport, Iowa; Taber & Co. of Keokuk, Iowa; W. and J. Fleming of McGregor, Iowa; B. Hershey of Muscatine, Iowa; Hemenway, Wood & Co. of Lansing, Iowa; Weyerhaeuser and Denckmann of Rock Island, Illinois; Keator and Wilson, and Dimock, Gould & Co. of Moline, Illinois; Hill, Lemon & Co., H. S. Parker & Co., J. Tamor & Co., and A. Baeckeler of St. Louis, Missouri; Youman Brothers and Laird, Norton & Co., of Winona, Minnesota. Hotchkiss, *Lumber and Forest Industry,* 635.

[12] Norton, *Mississippi River Logging Company,* 14–16, 34; *Chippewa Herald,* January 13, June 3, 1871; *Eau Claire Free Press,* May 11, 1871; William Blackstone, *Commentaries on the Laws of England in Four Books,* edited by Thomas M. Cooley (2 vols., Chicago, 1884), book 3, p. 4n; book 4, p. 167.

Lawrence, and the carrying places between the same, shall be common highways, and forever free" to all inhabitants of the territory and citizens of the United States. After 1789 the power of the federal government to regulate inland navigation was confirmed by judicial construction of the interstate commerce clause of the Constitution, but the states continued to hold a residual control over bordering and enclosed waterways. Thus the constitution of Wisconsin, approved in 1848, provided for "concurrent jurisdiction" over contiguous rivers, and repeated verbatim the free navigation clause of the Northwest Ordinance. The federal Supreme Court sought to untangle this potential conflict of jurisdiction by construing the control of the federal government as paramount but inoperative without specific congressional pronouncement; in other words, until Congress exercised its authority, state laws on the subject could not be contested.[13]

In 1871 Congress had not yet done anything to regulate navigation on the Chippewa. It appeared, therefore, that to put the Weyerhaeuser syndicate out of business the local lumbermen had only to induce the comparatively amenable state legislature to outlaw the works of the boom company as hazards to navigation. With this objective the Eau Claire group launched a rather haphazard campaign in 1871. To prove that Beef Slough was navigable in the ordinary sense of the term they appointed a

[13] See Ordinance of 1787, art. IV, sec. 14, in *Revised Statutes of the United States,* 1878, p. 16. The free-navigation clause of the Ordinance was incorporated in the constitution of Wisconsin in article XV, section 1. John Marshall wrote in 1824, "All America understands... the word 'commerce,' to comprehend navigation.... The power over commerce, including navigation, was one of the primary objects for which the people of America adopted their government and must have been contemplated in forming it." Gibbons *v.* Ogden, 9 Wheaton 190 (1824). Chief Justice Roger B. Taney went so far as to permit the extension of federal maritime jurisdiction to navigable inland waterways. Propeller Genesee Chief *et al. v.* Fitzhugh *et al.,* 12 Howard 443 (1851). The leading cases in which the court passed on the question of jurisdiction are Willson *et al. v.* Black Bird Creek Marsh Company, 2 Peters 245 (1829); State of Pennsylvania *v.* Wheeling and Belmont Bridge Company *et al.,* 13 Howard 518 (1851); and Gilman *v.* Philadelphia, 3 Wallace 713 (1865). The principle of paramount federal jurisdiction was applied directly to the Chippewa River in Pound *v.* Turck, 95 U. S. 459 (1877). The Wisconsin Supreme Court followed it in Wisconsin River Improvement Company *v.* Manson, 43 Wis. 255 (1877).

committee, which after a solemn investigation reported that the sheer booms of the logging company were diverting so much water that the navigability of the main channel was seriously diminished, whereas that of the slough was increased. Their most spectacular stunt was to run a lumber raft into the Beef Slough log reservoir to show how legal navigation was being impeded there. The campaign backfired, however. During the legislative session of 1872 the logging company, aided by assemblymen from eastern Wisconsin, obtained the repeal of all laws requiring that a passage be kept open in sloughs when there was a navigable channel open in the main part of the stream.[14]

The doctrine of free navigation having failed them, the local millowners tried a new tack. Four years earlier, in 1868, Levi Pond and the Eau Claire Lumber Company, his employer, had been granted a patent on the fin boom, the ingenious device by which logs were sheered from midstream into sorting and rafting works without preventing the passage of other traffic in the navigable channel. To the Beef Slough Company, which used a boom three thousand feet long to direct logs and water into the slough, such a contrivance was indispensable.

Pond's boom had been used for several years before 1868, and it is doubtful whether the courts would have upheld the patent he obtained, since federal law at that time disallowed the issuance of a patent for an article that had been in use longer than two years. The patentees, therefore, when the struggle with the Mississippi River lumbermen began to become serious in 1872, obtained the passage of a bill in Congress confirming their

[14] A. E. Pound to Ingram and Kennedy, April 18, 1871, Joseph G. Thorp to Daniel Shaw, January 18, 1872, and Thorp to Orrin H. Ingram, January 24, 1872, in the Ingram Papers; *Assembly Journal,* 1872, Appendix, 105; *Chippewa Herald,* January 13, June 3, August 5, November 25, December 9, 1871, January 27, March 9, 23, 30, June 29, 1872, April 12, 1873; *Eau Claire Free Press,* May 11, June 1, August 10, October 5, 1871, March 7, 28, 1872; *Chippewa Falls Democrat,* September 14, 1871; Oshkosh *Northwestern,* January 11, 1872; Norton, *Mississippi River Logging Company,* 35–36. There could have been no real doubt of the navigability of the main channel, at least as far as the falls. In 1867 six steamers in constant use transported five thousand passengers in addition to commodities and merchandise. The slough, however, had never been used for regular shipping, and the logging company could logically contend that it was not interfering with navigation. *Assembly Journal,* 1868, p. 630.

rights in the fin boom and making them retroactive to June, 1870. Realizing their vulnerability, the Weyerhaeuser associates fought vigorously but unsuccessfully to get the confirmatory law repealed. Early in 1874 the patentees brought suit against the Beef Slough and Mississippi River companies for infringement, asking damages of forty thousand dollars. After many dreary delays the plaintiffs won their case, but by that time they had already agreed to allow the defendants to use the boom for the life of the patent.[15] Why the Eau Claire Lumber Company should thus have relinquished an opportunity to deprive the Mississippi River lumbermen of an indispensable device has never been satisfactorily explained, but presumably the lure of fat royalties for the moment outweighed local patriotism. At any rate the Chippewa group had come to the end of another legal avenue.

After some months of hopeless drifting, during which they made a fruitless effort to strengthen the state laws forbidding obstruction of navigation, the local millowners decided to exploit the doctrine of free navigation once more, this time in the federal courts. In coming to this decision they were influenced by the fact that a federal appropriation had been made for a survey of the Chippewa River in 1876, which injected a new element into the situation. The state Supreme Court had held in 1868 that any stream capable of floating logs to market was by common law a public highway and that a logger could use his own devices for getting past obstructions so long as he caused no unnecessary damage. This seemed to imply that a court order might be obtained to remove obstacles to navigation. But since the obstructions on the Chippewa to which the local lumbermen objected were certainly authorized by the legislature, it was doubtful whether any state court would order their abatement. On the other hand, the federal Supreme Court had decided that congressional legislation could at any time supersede state laws

[15] *Sources of the Mississippi, Saint Croix, Chippewa, and Wisconsin Rivers* (45 Congress, 3 session, House Executive Document no. 54, serial 1858), 8; *Eau Claire Free Press,* March 12, 1874, January 28, 1875, January 13, 1876; La Crosse *Republican and Leader,* March 21, 1874; *Chippewa Herald,* June 19, 1874, February 5, 1875; *United States Statutes at Large,* 17:389; *United States Senate Journal,* 43 Congress, 1 session (1873–74, serial 1579), 163, 319; *House Journal,* 43 Congress, 1 session (1873–74, serial 1593), 279.

on the subject. The survey appropriation, it could be argued, constituted an assumption of federal authority on the river.[16]

Accordingly, pleading the paramount authority of the federal government, Thaddeus C. Pound of Chippewa Falls early in 1877 organized a movement to induce the Department of Justice to seek an injunction abating the works of the Beef Slough Company and its affiliates and preventing them from running logs in the Chippewa River. In April Attorney General Charles Devens appointed a special assistant to start proceedings in the Circuit Court of Western Wisconsin, but political pressure delayed the institution of the suit until midsummer.[17]

In its bill of complaint the government alleged that the works of the Beef Slough Company obstructed navigation in the main channel of the river and in the slough, and that its booms deflected so much water from the main channel that navigation was thereby seriously impaired. "Without warrant of law"[18] the logging company was driving logs from the upper Chippewa in such numbers that steamboats, ferries, and lumber rafts were virtually being driven from the river. The logs belonging to the mills on the Chippewa were being carried along to Beef Slough; and when the mills made an effort to detain all the logs until their own could be segregated, the logging company, using "large gangs or bodies of men employed by them therefor, cut and opened all booms along said river . . . and thus cast loose

[16] H. P. Graham to Orrin H. Ingram, February 11, 12, 1875, March 6, 1876, in the Ingram Papers; Whisler *v.* Wilkinson *et al.*, 22 Wis. 546 (1868). In three cases tried during 1877 the state Supreme Court, while restating the principle that any stream capable of floating logs was legally "navigable," affirmed the right of the legislature to regulate their use for logging. Tewksbury *v.* Schulenberg *et al.*, 41 Wis. 584; Olson *v.* Merrill, 42 Wis. 203; and Wisconsin River Improvement Company *v.* Manson, 43 Wis. 255. For the United States Supreme Court's reaffirmation of the supremacy of federal jurisdiction see the Chippewa River case, Pound *v.* Turck, in 95 U. S. 459.

[17] Letters to Orrin H. Ingram from H. H. Douglas, April 29, 1877, H. L. Humphrey, July 6, 11, August 5, 1877, and Thaddeus C. Pound, August 29, 1877, from E. R. Leland to Clarence Chamberlin, May 9, 1877, from Thaddeus C. Pound to Ingram and others, July 7, 1877, and from T. C. Rogers to Ingram and Kennedy, August 2, 1877, all in the Ingram Papers; *Eau Claire Free Press,* January 25, April 5, 26, May 3, July 19, August 2, 9, 1877; *Chippewa Herald,* January 26, April 13, 20, 27, July 6, 13, 20, 27, August 3, 10, September 28, 1877.

[18] The company's charter did not specifically authorize the driving of logs from upriver, but only the handling of them in the slough.

and drove down said river not only their own but all the logs of other owners detained in said booms, and . . . by force and violence and by riotous and unlawful conduct and personal injury to owners objecting to the same, so repeatedly opened booms, and forced . . . great numbers of logs belonging to other persons . . . into said Beef Slough, and then . . . charged and exacted . . . unlawful tolls, charges and tariffs." So long as the logging company continued thus to obstruct traffic, the sums spent by the United States government on improving the river were wasted. Finally, none of these acts of the company was necessary, for log rafts could be assembled above the head of navigation and floated down the river under full control.[19]

After more than a year of litigation the court, following the precedents of earlier decisions, sustained the demurrer of the defendants. Admitting that "undoubtedly Congress has the power to legislate as to all rivers navigable in fact," the court could find no federal law specifically defining how logging works could be constructed on the Chippewa River. In the absence of such congressional legislation the regulation of a navigable stream was left entirely to the state; the federal government could not in equity abate any obstruction. The fact that Congress had appropriated money for a survey of the Chippewa River did not affect the issue. The Beef Slough Company had not interfered with the survey, and the court held that it had no power to intervene unless the operation of the federal law had actually been prevented.[20]

When it became apparent that the case would be lost, the Chippewa millowners fell back upon their last legal resort. Captain Edward E. Heerman, a Minnesotan who owned a line of steamboats on the Chippewa, was induced to start two actions in the federal courts against the logging company: one a suit for damages sustained by his boats in colliding with loose logs, the other a plea for an injunction to prevent the running of such logs to the detriment of steamboating. In October, 1879, the

[19] *Information and Bill of Complaint . . . U. S. v. Beef Slough . . . Company.*

[20] United States *v.* Beef Slough Manufacturing, Booming, Log-Driving, and Transportation Co. *et al.*, 24 *Federal Cases* 1064 (1879).

circuit court handed down its decision in the injunction case, the more important of the two.

No court of equity, the decision maintained, could at the instance of a private individual abate as nuisances the works of a private corporation in a river lying wholly within a single state if their construction had been authorized by the legislature. Even if the corporation had exceeded its powers and their works did actually obstruct navigation, a court of equity could not interfere until the rights of the complainant had been established by law. The interests involved were "very large," and the alleged nuisances had been "acquiesced in for a long time." Certainly the logging company had the right, common to all, to use the Chippewa as a public highway for the transportation of logs and lumber. That right would continue to exist until it was "changed or abrogated by the legislature of the state or by Congress." True, the commerce clause of the Constitution empowered Congress to subordinate logging to steamboat interests, but until it did so the state had full power to legislate.[21]

The Mississippi River lumbermen were of course jubilant. No longer could it be argued, one of them wrote, that "God made the river only for steamboats."[22] They overlooked the significant suggestion of the court that Congress might some day subordinate their interests to those of steamboat transportation.

Despite the best efforts of the Chippewa millowners the Mississippi River Logging Company had grown and prospered. The franchise of the Beef Slough Company gave it virtually complete control of logging operations at the mouth of the Chippewa. Weyerhaeuser had in 1871 transferred his lease of the Beef Slough Company's works to the Mississippi River Logging Company, which shortly thereafter had bought five hundred and two of the thousand shares of Beef Slough stock. In 1881 the Weyerhaeuser group purchased the remainder of it. From time to time the Wisconsin legislature amended the Beef Slough charter to

[21] Edward E. Heerman to Orrin H. Ingram, June 30, 1878, in the Ingram Papers; *Chippewa Herald,* March 7, October 10, December 19, 26, 1879; Oshkosh *Northwestern,* October 9, 1879; *Eau Claire Free Press,* March 6, 1879; Heerman *v.* Beef Slough Manufacturing . . . Co. *et al.,* 1 *Federal Reporter* 145 (1880).

[22] Norton, *Mississippi River Logging Company,* 17.

increase the powers of the company, which by 1879 was employing four hundred men in its rafting works alone. In 1880 the syndicate obtained from Congress, possibly in view of the court decisions of 1879, a franchise to build sheer booms in the Mississippi and other works in Rollingstone or Straight Slough, on the Minnesota side below the mouth of Beef Slough. This law, ostensibly providing a means of catching logs that had escaped from the Beef Slough works into the Mississippi, later became the basis for much more extensive operations.[23]

The Mississippi River Logging Company had become the greatest financial power on the upper river, but Weyerhaeuser, who had been elected its president in 1872, realized that to establish the organization on a firm basis it must buy timberlands. In 1876 the company purchased from an agency in Ithaca, New York, fifty thousand acres originally owned by Cornell University, and from then on it added regularly to its holdings, buying from the land-grant railroads and other large holders. So vast had the timber empire of the Weyerhaeuser syndicate become by the nineties that it was widely believed to monopolize the raw materials of the entire lumber industry of the upper Mississippi.[24]

The baffled Chippewa lumbermen, helpless to prevent the driving of logs down the Mississippi and seeing the Weyerhaeuser group increasing its power, were ready by 1879 for conciliation. For their part the Mississippi River lumbermen, suffering from discriminatory railroad rates, were equally willing to come to some sort of terms. Members of both groups had always

[23] *Sheer-Booms in the Mississippi River* (46 Congress, 2 session, House Report no. 664, serial 1936, Washington, 1880); *United State Statutes at Large,* 21:282; *Eau Claire Free Press,* February 12, 1880; Norton, *Mississippi River Logging Company,* 38, 48; *Northwestern Lumberman,* June 7, 1879.

[24] It was Weyerhaeuser's custom to buy timber with his own money whenever he found a bargain and sell it to the company at cost. A member of the syndicate once asserted that the company never bought land at the government price of a dollar and a quarter an acre, but paid an average of ten to twenty dollars an acre. Norton, *Mississippi River Logging Company,* 43, 45–46, 51, 60, 92–94; La Crosse *Chronicle,* November 18, 1880; *Eau Claire Free Press,* October 25, 1875, February 2, 1893; Paul W. Gates, *The Wisconsin Pine Lands of Cornell University: A Study in Land Policy and Absentee Ownership* (Ithaca, New York, 1943), 217.

had close business relations even at the height of the conflict, and they now found it easy to effect an agreement. In the spring of 1879 the Mississippi River Logging Company made arrangements with the Chippewa River Manufacturers' Driving Company to conduct a joint drive down the Chippewa, and a little later Weyerhaeuser began to operate the flooding-dam facilities controlled by his company in a fashion that was advantageous to the lumbermen in the Eau Claire area. In 1880 the two groups evolved a mutually satisfactory system of log exchange.[25]

Finally, in December of 1880, the Chippewa River lumbermen and representatives of the Weyerhaeuser syndicate met at Chicago for three days and ironed out all their differences. It was agreed that the two groups should pool their interests on a proportional basis. Several common agencies would be set up to cut, drive, and saw logs, the Mississippi River group receiving sixty-five per cent of the profits of these activities. The first step was the purchase of the Chippewa Lumber and Boom Company at Chippewa Falls, with all its timber holdings and uncut logs. Sixty-five per cent of its stock, in accordance with the agreement, went to the Mississippi River Logging Company and its members; the remainder was divided among the Chippewa lumbermen. To operate the "big mill" at Chippewa Falls and to make all future purchases of pinelands, the lumbermen organized jointly, in September of 1881, the Chippewa Logging Company, which came to be known as "the pool," the stock being distributed on the same basis as that of the boom company. At the same time the syndicate, so that it might control all the organizations buying and handling logs on the river, acquired the stock of the Chippewa River Improvement and Log Driving Company and sold thirty-five per cent of it to the local lumbermen. A complete community of interests had now been established, and the two groups began to work together on extremely amicable terms.[26]

[25] *Chippewa Herald,* July 11, 1879, March 5, July 16, 1880; *Northwestern Lumberman,* April 19, 1879; Norton, *Mississippi River Logging Company,* 17–18, 52.

[26] Thomas Irvine to Orrin H. Ingram, August 30, 1881, and Frederick Weyer-

In the years thereafter the syndicate expanded its activities in many directions; it built logging railroads, enlarged the capacity of its mills, and bought more mills and timberland. In 1889 a deposit of sand at the head of Beef Slough threatened to render useless the company's works there. The federal government refused to improve the old channel east of the head of the slough and likewise denied the company permission to place piers below the head of the slough to force logs into the booms. Weyerhaeuser and his associates then decided to abandon Beef Slough and remove their sorting works to West Newton Slough, about seven miles below the mouth of Beef Slough on the Minnesota side of the river. This location necessitated the driving of logs across the Mississippi, a navigable river, but the company disregarded the looming opposition of navigation interests and organized the Minnesota Boom Company of Winona, which built improvements at West Newton.[27]

The removal of the logging company's rafting works again aroused the ire of those opposed to shipment of raw materials out of Wisconsin. In 1891 a member from Chippewa County introduced in the Assembly a bill prohibiting the running of loose logs on the Chippewa below the head of Beef Slough. As passed, however, the measure merely provided for the erection of dams and sorting works between the flowage of the Eau Claire dam and the city of Chippewa Falls. Apparently the sponsors of the amended bill hoped that adequate sorting and rafting works in the Chippewa would induce the syndicate to give up its project at West Newton. In the same session the legislature imposed on all dam and boom companies deriving their authority from the state an annual license fee of two per cent of their gross earnings. The sponsor of this law, an assemblyman from Green County, asserted that it was directed against the "foreign" corpo-

haeuser to Ingram, December 27, 1884, in the Ingram Papers; *Chippewa Herald,* December 10, 1880, April 8, 1881; La Crosse *Chronicle,* April 14, 1881; Norton, *Mississippi River Logging Company,* 30–31, 50, 55–59; Hotchkiss, *Lumber and Forest Industry,* 635.

[27] *Eau Claire Free Press,* January 18, 1883, September 22, 1887, April 25, May 21, August 27, 1889, January 23, 30, February 6, 1890; *Chippewa Herald,* September 22, 1882, July 20, 1883, August 14, 1885, July 23, 1886, October 14, 1887, August 17, 1888, January 3, 1890; La Crosse *Weekly Chronicle,* April 18, 1889; Norton,

rations who were taking materials from the state without paying adequate taxes.[28]

Navigation interests on the Mississippi were also much displeased with Weyerhaeuser's latest strategem. The city of Wabasha was particularly indignant because the logs driven across the river interfered with the operation of its ferry. Weyerhaeuser sought to silence this opposition by obtaining congressional sanction of the Minnesota Boom Company's transportation of logs across the Mississippi. A bill to that effect, introduced by William D. Washburn of Minnesota, was passed by the Senate in February, 1891, but a similar House bill, sponsored by Mark H. Dunnell, was buried in the Commerce Committee, despite the fact that spokesmen for Weyerhaeuser had induced Wabasha to withdraw its objection in return for a promise that he would build a bridge and make provision for the operation of the ferry "at least part of the time." The death of the measure was due largely to the opposition of the legislature of Wisconsin, which sent a memorial to Congress violently condemning it.[29]

Despite his failure to get congressional authorization, Weyerhaeuser continued to move logs across the river. He could still rely on the law of 1880 authorizing his company to build sheer booms on the Mississippi River "for the purpose of sheering logs that may escape and float out of the Chippewa River" from the main channel into Rollingstone or Straight Slough. This law and others, such as the Rivers and Harbors Act of 1890, entrusted the defense of navigation rights to the secretary of war, who was given a fairly free hand in the performance of this duty. He did not choose to interfere with the Weyerhaeuser interests.[30]

Mississippi River Logging Company, 68–75; Hotchkiss, *Lumber and Forest Industry*, 635.

[28] *Assembly Journal*, 1891, pp. 266, 998, 1022, 1121; *Session Laws*, 1891, ch. 313, 422; *Eau Claire Free Press*, February 26, April 16, 30, May 7, 1891; *Chippewa Herald*, April 3, 1891.

[29] *Chippewa Herald*, May 31, 1889, June 6, 1890, September 3, 1897, May 25, 1900; La Crosse *Weekly Chronicle*, February 12, 1891; *Eau Claire Free Press*, February 19, April 30, 1891, March 3, 10, 24, 31, 1892, July 20, 1893, June 28, 1894, July 8, 1897; *Congressional Record*, 51 Congress, 2 session (1890–91), 1892, 1947, 2368, 2427.

[30] *United States Statutes at Large*, 21:282; 26:426.

The inactivity of the War Department engendered angry protests in the Chippewa region. In 1892, when the promised bridge had not yet materialized, the irate city of Wabasha sent photographs to Washington showing the Father of Waters choked with Weyerhaeuser logs. Within a year a pontoon bridge was built at Wabasha, but floating logs made its operation uncertain. Boat owners, ferry riders, and farmers "lingered at the shore and damned Weyerhaeuser and his logs." More serious for the syndicate was the allegation made in a memorial of the Wisconsin legislature to the House of Representatives that a government dam had been blown up to facilitate the West Newton operation. The House passed a resolution calling for information, which elicited reports of the episode from the War and Justice departments, as well as explanations why navigation on the Mississippi continued to be obstructed.

A dam at the head of West Newton Slough had indeed been blown up in 1889 by employees of the Mississippi River Logging Company, reported Major Alexander Mackenzie of the Rock Island office of Army Engineers. In accordance with provisions of the Rivers and Harbors Act of 1876 the matter had been taken before a grand jury at La Crosse, but the case had been thrown out on the ground that the jury lacked jurisdiction over an offense committed in Minnesota. The district attorney at La Crosse had thought further grand jury action to be inadvisable because the penalty for the offense was negligible. Rather he advised that an injunction be issued to prevent further driving of loose logs into the Mississippi. Agreeing that an indictment would accomplish little, Mackenzie had argued that the injunction was equally inexpedient. The conflicting navigation interests, he thought, should settle their differences without help from the government. When in April, 1891, the United States attorney for the district in Minnesota had asked what to do about the case, Mackenzie had advised him to drop it. The Mississippi River Logging Company was undoubtedly disturbing navigation by running its logs in the river, but the company claimed that it was itself a navigation interest with the same rights as any other. He did not, he said, want any special legislation for the regulation of the river; the courts could settle the dispute. As a matter of fact, he concluded, the Rivers and Harbors Act of 1890 gave the War Department ample authority to investigate and punish the syndicate if its works were detrimental to navigation, but for the present it was not planning to take any action against the company.[31]

This complaisant attitude of its Rock Island office governed the policy of the Army Engineers, and for nearly ten years longer Congress continued to receive protests from annoyed river communities. At last the legislators felt constrained to act. The Rivers and Harbors Act of 1899 forbade unequivocally the floating of loose logs or sack rafts "in such a manner as to obstruct, impede, or endanger navigation" in streams "actually navigated by steamboats." Faced with complete defeat, the Weyerhaeuser syndicate mustered enough influence to induce the House a year later to pass a bill exempting the St. Croix, the Chippewa, and portions of the Mississippi from the prohibition, subject to regulations made by the secretary of war. The Senate amended the House bill in accordance with the recommendation of Major Mackenzie, then acting chief of engineers, to make it applicable to all the navigable streams in the country. As enacted into law the bill exempted any stream on which the floating of loose logs and sack rafts was the chief form of navigation. For such waterways the secretary of war was ordered to make rules that should take into account the conflicting navigation interests. When these regulations had been properly published, they should have the force of law.[32]

The struggle thus ended in a working compromise, essentially fair but not wholly satisfactory to either side. Nevertheless the friendliness of the War Department enabled the logging syndicate to continue its operations until the pending exhaustion of

[31] *Congressional Record,* 52 Congress, 1 session (1891–92), 1704; *Eau Claire Free Press,* March 10, 24, 1892, July 20, 1893; *Government Dams on the Mississippi River* (52 Congress, 1 session, House Executive Documents nos. 178, 183, serial 2955, Washington, 1892).

[32] *United States Statutes at Large,* 30:1152; 31:172; *Loose Logs, Steamboats, and Rafts on Certain Rivers and Streams* (56 Congress, 1 session, Senate Report no. 1035, serial 3890, Washington, 1900); *Chippewa Herald,* March 2, 23, 1900.

Chippewa pine timber forced the closing of the West Newton works in 1909.

The logging companies controlled by the Mississippi River lumbermen absorbed an enormous amount of Chippewa valley timber. From about twelve million board feet of logs in 1867 the annual traffic passing through Beef Slough increased to a maximum of more than five hundred million feet in 1885. The total for the period 1867–89 exceeded five billion board feet. The West Newton works reached their maximum activity in 1892, when they handled six hundred million board feet. The vast scale of its operations enabled the organization founded in 1871 to swallow up or dictate to virtually all lumber manufacturers on the Chippewa, St. Croix, and upper Mississippi rivers.[33]

In 1900 the Weyerhaeuser companies still held enormous amounts of land in Wisconsin and Minnesota. One of them, the Flambeau Land Company, owned a hundred and sixty thousand acres of cutover lands in Chippewa and Sawyer counties. In 1907 the Weyerhaeuser associates created the American Immigration Company to handle and sell these lands. With the exhaustion of Chippewa River pine an accomplished fact, the officers of the Mississippi River Logging Company were authorized to dispose of all remaining assets of the company. In 1910 the Mississippi River Logging Company of Chippewa Falls was organized to sell whatever property had not been liquidated. The Midwestern activities of Weyerhaeuser and his syndicate had come to an end with the epoch in which they had risen to power and wealth. During that epoch nearly every one of his close associates had become a millionaire.[34]

[33] Hotchkiss, *Lumber and Forest Industry,* 635–636; Usher, *Wisconsin,* 1:194; *Chippewa Herald,* January 5, 1900.

[34] *Chippewa Herald,* January 5, June 15, 1900; Norton, *Mississippi River Logging Company,* 90; Blair, *Raft Pilot's Log,* 54.

10. The Land Problem

ALTHOUGH the lumbermen actually needed, and usually really wanted, only the timber on a forested tract, under the public-land system of both the federal and the state government they had no choice but to buy the land itself. The growth of the industry therefore led to the rapid conveyance of the public timberlands to private hands. Of the original thirty-five million acres of public land in Wisconsin, only four thousand acres remained unappropriated in 1924. By 1932 ninety-two per cent of the existing forest acreage in Wisconsin was privately owned. What happened in Wisconsin was typical of all the forested regions of the country. The federal Bureau of Corporations estimated in 1911 that during the preceding forty years the timberland to which the government held title had decreased from three-fourths of the available timber to one-fifth.

In Wisconsin by far the greater part of the timberlands passed from the federal government to private owners either directly through public sale, pre-emption, and free homestead grants, or indirectly by way of grants to the state for special purposes of one kind and another. The land law of 1820, which remained in force until 1889, permitted the direct sale of eighty acres or more for a minimum price of a dollar and a quarter an acre. All land offered for sale under this law was first presented at public auction, usually for a period of two weeks, and only those lands not sold at the auction could be bought at private sale for the minimum price. The Pre-emption Act of 1841 allowed a person who had "squatted" on unsurveyed public land and improved it to buy a farm up to a hundred and sixty acres at the minimum price before the land was offered to the public. Although this law was not repealed until 1891, it was theoretically superseded by the Homestead Act of 1862, under which any family head, widow, or man who had reached the age of twenty-one could acquire a quarter section for a merely nominal fee if he had lived on the land and improved it for five years. In addition to these laws, under which the government disposed of immense quantities of timberlands, there were many special grants to the state in aid of railroad, canal, or wagon-road construction, public education, and the improvement of swamp lands.

The primary purpose of the land laws was to foster agriculture and to produce a nation of small farmers. They were wholly inapplicable to timberlands or any other lands valuable for non-agricultural resources. But with few exceptions Congress ignored that fact. Not until 1878 was an act passed to govern specifically the disposition of government-owned forest land. This law, the Timber and Stone Act, applied originally only to the Pacific coast and Nevada, and it changed the direct-sale provisions of the old law only by raising the minimum price to two dollars and a half an acre and by increasing to a quarter section the minimum sale of land valuable for its timber. Although its provisions were extended to all public-land states in 1892, it had slight significance for the Wisconsin lumber industry, since there remained in the state at that time less than two hundred thousand acres of unappropriated federal land.[1]

The lumbermen of Wisconsin were therefore forced to acquire their timberland under a system that in no way met their needs. They cared little for the land itself, but since they had to buy it, they would have preferred it in large blocks. It was not economical to acquire the right to cut timber from many scat-

[1] *Report of the Commissioner of the General Land Office,* 1924, p. 86; [Wisconsin] Committee on Land Use and Forestry, *Forest Land Use in Wisconsin,* 14; Bureau of Corporations, *The Lumber Industry,* pt. 1, pp. xvii, 182; *United States Statutes at Large,* 3:556; 5:455; 12:392; 20:89; 28:348.

tered patches of eighty or a hundred and sixty acres. For them the ideal system would have been one under which they could have bought the right to cut trees from large tracts without having to buy and pay taxes on the land.

Since no such system was instituted during the heyday of the lumber industry in Wisconsin, the lumbermen made the best of the existing laws. The land system that was intended to build up many small farms was used in timbered regions to build up large individual holdings. Such an application of laws designed for an entirely different purpose led inevitably to fraud, and it fostered the most unfortunate features of land speculation. Because the state and private corporations which had received large grants of timberland from the government perpetuated the shortcomings of the federal land system, the disposition of these lands, too, was sometimes attended by dishonesty. A veteran logger characterized the transition from public to private ownership as "one of the greatest periods of graft and exploitation . . . that this nation has ever gone through. The rule was to beat the other fellow before he beat you and it was followed with a conscientiousness born of greed and avarice."[2]

The attitude of the West toward forest land and the peculiar nature of the lumber industry did much to foster dishonesty in the acquisition of timberlands and ruthless destruction of their resources. The average Westerner of the nineteenth century had an inordinately possessive attitude toward the public lands in his area. He wanted no control or regulation from a distant government in the disposition of them. "The public lands of the country," wrote a Wisconsin newspaper editor in 1871, "belong to the people as much as the sun shine and the air and the waters of our lakes and rivers, and the government should simply regulate their disposal so as to secure every man desiring their use, in his actual occupation and cultivation."[3] The frontiersman wanted the public lands to come into private possession as quickly as possible so that they might be "improved." Land in itself did not represent wealth to him; there was too

[2] Nelligan, *Life of a Lumberman,* 175, See below, Chapter 11.
[3] Oshkosh *Northwestern,* February 16, 1871.

much of it. Growing trees were only an obstacle to progress; cut and sawed, they furnished raw materials for an industry that created employment and drew wealth into his home area. The clearing of the forests, he thought, also promoted agriculture, the inseparable companion of "progress." To him the best land policy was one that quickly transformed the wilderness into valuable private property.

To the lumberman, with his special needs, this attitude was a godsend. It enabled him to carry on his business for many years under a land system that would otherwise have been a serious handicap. Utilized to the fullest, it gave him the best opportunity he could have had to make a great fortune—by purchasing at a low price great stretches of timberland that steadily increased in value.[4]

Before buying timberland the lumberman or speculator usually had it "looked up" by an expert woodsman. Men who made it their trade to examine forest land for others were known as "landlookers" or "timber cruisers." Some were in the employ of a lumber or land company; others were independent businessmen who sold their information to the highest bidder. Ordinarily the independent timber cruiser also had some other occupation, such as running a logging crew, scaling timber, or guiding prospective settlers and sportsmen. In the lake states the estimation of timberlands became a highly skilled trade. As a class, according to a member of Wisconsin's first Forestry Commission, the cruisers had "remarkable intelligence" and a "great stock of empirical information regarding such matters as fall within the immediate scope of their business."[5]

Timber cruisers usually worked in groups of three or five. Traveling overland or by river to the area in which they were to work, they established a base camp and supply depot. Then each man would set out for himself to inspect designated sec-

[4] Stephenson, *Recollections*, 63–64; Oshkosh *Weekly Northwestern*, September 17, 1885.

[5] Bruncken, *Forests and Forestry*, 81–82; A. M. Dole to Ingram and Kennedy, January 9, 1858, in the Ingram Papers; *Wisconsin Lumberman*, 1:200 (January, 1874); Woodward, *Valuation of American Timberlands*, 36. The Holt Lumber Company Papers contain cruisers' plat books and reports.

tions of land, the exact location of which could easily be discovered by using the field notes of the government surveyor, the survey markings, and a compass. Having found one corner of a section or any other unit, an experienced cruiser could pace off the boundaries with astonishing accuracy. In heavily wooded areas five hundred paces were equal to eighty rods, two thousand paces to a mile; in open hardwood country eighteen hundred paces were the equivalent of a mile. As soon as he had located the boundaries the cruiser, if he was alone, paced off a two-hundred-foot square within the tract, estimated or counted the number of trees within the square, and scaled a few of them. The average number of board feet per tree was multiplied by the average number of trees to the acre, and this product was multiplied by six hundred and forty to obtain the approximate number of board feet in a section. When two or three men examined a section together, they worked through it in such a way that nearly every part of the land was seen.

To be efficient a timber cruiser needed keen judgment and wide experience. Trees could be very deceptive; sometimes, for instance, a tree with a healthy bark was hollow inside. After several years at his trade a landlooker became so expert at judging distances and timber footage that he could climb a tree and from this vantage point locate and estimate with fair accuracy the timber in a large area. According to one old woodsman, "he daily experiences hardships such as working his way up rivers with many swift waters, and crossing lakes in birch-bark canoes, in wind storms and in rain; fording streams . . . ; building rafts to cross rivers and lakes; climbing through windfalls; crossing miles of swamp . . . ; and traveling sometimes all day in an open and burnt country with his bed and board upon his back, the sun's hot rays pressing like a heavy weight upon his head, while myriads of black flies swarm about him and attack every exposed inch of his skin." But there were compensations in good health, buoyancy of spirit, and opportunity to enjoy the "solitude of the great unbroken forest."[6]

[6] George H. Warren, *The Pioneer Woodsman As He Is Related to Lumbering in the Northwest* (Minneapolis, 1914), 47–48; Nelligan, *Life of a Lumberman,*

In the woods there was much rivalry between landlookers working for different employers. When by chance they met, they engaged in a contest of wits to reach the land office first, for the labor of weeks would be wasted if a rival entered the land on which a cruiser had been working. It is surprising how few timber cruisers made use of their information to grow rich themselves by speculation. It may be, as a forestry expert has suggested, that "their constant life with unsophisticated nature kept them from acquiring that worldly shrewdness which is indispensable for success in money-making."

This is not to say that there were no dishonest landlookers. Their pay, though it compared favorably with that of skilled labor in other branches of the industry, was not high enough to protect a weak character from accepting a bribe to report valuable land as worthless and thus enabling an unscrupulous competitor to buy it cheap. Occasionally a landlooker in the employ of a lumber company would falsify his report so that he himself could enter the land. One rich speculator confessed in 1894 how he had got his start: "One day I was traveling through the woods locating land when the thought struck me that I was a pretty considerable big d—n fool. Why should I locate all the land for the company and none for myself? I made up my mind to be a d—d fool no longer." Accordingly he marked his platbook in such a way that "a lot of land" containing timber appeared to be swamp. His company failed to buy the land, and he took it for himself for six hundred dollars, which he borrowed. Before the year was out he had sold part of it for seventy-five hundred dollars and had cut nine million feet of timber from the rest. "I didn't work for the company after that," he added gratuitously.[7]

It was certainly not ignorance of land values that prevented Congress from enacting adequate regulations for the disposition of timberland in Wisconsin. Repeatedly the commissioners of the General Land Office reported that there was a wide difference between the price the government set on its pinelands and

49–51; Augustus Cole to Holt and Balcom, April 15, 1887, in the Holt Lumber Company Papers.

[7] Warren, *Pioneer Woodsman*, 49; Bruncken, *Forests and Forestry*, 82; *Chicago Times*, May 26, 1894.

their actual market value. Theoretically the price of a dollar and a quarter an acre set by the government (two dollars and a half within the primary limits of railroad grants) was only a minimum. It was expected that at the public auction preceding private land sales something approaching the true market value would be obtained. The theory did not work out in practice, however. The "field notes" of the government surveyor did not indicate what lands were valuable for pine, and Congress made no other provision for discovering their real value. Accordingly the official in charge of the public sale had no way of knowing what kind of land he was selling.

Moreover, the purchasers attending public auctions worked out in collusion various ways of keeping the price down to a "merely nominal figure." The result was that even the public sales usually brought only the minimum price, and lands not sold at the auction immediately became subject to private purchase by speculators who could then take them for the minimum price. It was a simple matter for a lumberman who wanted to buy land at a low price to bid ridiculously high at the public sale and then quietly forfeit his purchase after the sale had closed, allowing a partner to buy it privately before news of the forfeiture had leaked out.[8]

If old anecdotes may be credited, some lumbermen, abetted by dishonest officials, were not above using actual force to buy their lands at the lowest figure:

> When a parcel of good timber land was to be sold at auction they merely sent to the spot a squad of trusty strong-arm lads from the logging camp.
>
> "How much is offered for this valuable land?" says the auctioneer. . . .
>
> "One dollar and twenty-five cents an acre," says the agent of the lumber magnate.
>
> "One dollar and fifty cents," says some one else.

[8] *Report of the Commissioner of the General Land Office*, 1874, p. 5; 1875, pp. 9–10; *Wisconsin Lumberman*, 1:5 (October, 1873); *Eau Claire Free Press*, August 28, 1873. See Richard N. Current, *Pine Logs and Politics: A Life of Philetus Sawyer* (Madison, 1950), page 23, for the story of a ruse worked by Sawyer to acquire valuable lands at a low price.

Immediately the strong-arm lads in wedge formation, place this intruder at their head and march him at double quick to the outside world and far away from troublesome scenes.

"Sold for one dollar and twenty-five cents an acre," says the auctioneer.[9]

Time and again the General Land Office recommended that Congress provide for a survey to determine the value of the country's pinelands and for their sale at a just evaluation. Congress responded with the inadequate Timber and Stone Act of 1878, and pineland continued to be sold at less than the market price until 1889, when cash entries were suspended altogether in every public-land state except Missouri.[10]

Lumbermen and speculators could also buy pinelands from the state government. Through a series of federal land grants made in the years 1838 to 1866 the state of Wisconsin came into possession of about five million acres, exclusive of railway and canal grants. These land subsidies were for the usual purpose of promoting education and internal improvements. The largest was the transfer to the state of all the swamp and overflowed lands within its boundaries, the proceeds from which were to be used for drainage projects. The swampland grant of 1850 and later additions to it gave the state about three million acres, a large part of which was covered with valuable timber.[11]

In disposing of these lands the state followed the plan which by 1850 had become the basic pattern of federal policy: direct sale at public auction with pre-emption rights for qualified occupants. The terms of the federal grant did not permit the state

[9] Russell, *A-Rafting on the Mississip'*, 48.

[10] *Report of the Commissioner of the General Land Office*, 1874, p. 5; 1875, p. 10; *Sale of Stone and Timber Lands* (52 Congress, 1 session, House Report no. 482, serial 3043, Washington, 1892); *Chippewa Herald*, March 15, 1889; *United States Statutes at Large*, 25:854.

[11] *Legislative and Congressional Acts Concerning the Land Grants by the United States to the State of Wisconsin* (Madison, [1878]); *Compilation of United States Laws, State Laws, Decisions, Messages, and Correspondence Relating to the Swamp and Overflowed Land Fund of the State of Wisconsin* (compiled by Ernst G. Timme, secretary of state, Madison, 1882); Bureau of Corporations, *The Lumber Industry*, pt. 1, pp. 225–226; *United States Statutes at Large*, 9:519; Daniel S. Durrie, "The Public Domain," in *History of La Crosse County, Wisconsin* (Chicago, 1881), 225; *Report of the Commissioners of School and University Lands of the State of Wisconsin*, 1865, p. 9.

to make free gifts of land to private corporations and individuals, a basic feature of federal policy after 1860, but the terms of sale and the minimum price were liberal enough to satisfy most recipients of the state bounty.

Although subject to constant revision, the act adopted by the legislature in 1849 continued to represent the basic land policy of the state. It entrusted the state lands to a special commission which from time to time might sell to the highest bidder any part of the school and university lands it saw fit to relinquish. No land could be sold, however, for less than the "appraised value" previously established by the commission. Once offered in vain at a public sale, the land could be sold privately to the first person offering to pay the last appraised value. The maximum sale to any one person was a hundred and sixty acres, and pre-emptors were allowed only forty acres at a dollar and a quarter an acre. Buyers could pay as little as ten per cent down, and were given ten years to pay the balance. Instead of making his annual payment the purchaser might withhold it "as a loan from year to year," paying interest on it to the state.[12]

The generous terms of the original statute and its further liberalization in response to public opinion enabled the state to dispose of its lands in large blocks at low prices. In 1855 a committee of the Assembly, charged with the task of planning legislation for the disposition of swamplands, recommended that they be sold in large blocks as quickly as possible to prevent, among other things, their being stripped of timber by trespassers. This principle had already been applied to the school lands; in 1854, for example, the state sold seventy thousand acres to a single individual. Four years later Governor Alexander Randall warned the legislature that such haste was allowing speculators to acquire most of the state lands, but his words had no effect.[13]

[12] *Session Laws,* 1849, ch. 212; *Revised Statutes,* 1849, ch. 24. The minimum price of university lands was fixed at ten dollars in 1850, but the generally established minimum after 1852 was a dollar and a quarter. After 1852 pre-emptors were allowed 160 acres if they had been using the land for "agricultural or manufacturing purposes." *Session Laws,* 1850, ch. 176; 1852, ch. 9, 124.

[13] *Report of Mr. M'Indoe, from the* [Assembly] *Committee to Whom Was*

The state sold its lands at a price far below the market value. Between 1855 and 1870 the average price of good pineland on the open market ranged from two and a half to five dollars an acre, and occasionally a sale brought as much as twenty dollars. Yet purchasers of state pinelands were paying only a dollar an acre or less. Frequent special laws reduced the price still more in certain areas. To curb the speculation that inevitably resulted Governor Lucius Fairchild recommended in 1871 that all state lands be withdrawn from the market and honestly appraised so that a just price could be demanded. This suggestion was not adopted until nearly the close of the century. Meanwhile the state lands continued to be sold for prices ranging from fifty cents to three dollars an acre, whereas lands of equivalent value sold regularly on the open market for fifteen dollars and more.[14]

Low prices and sales in large blocks had within a generation reduced the state's land reserve by nearly one-half. By 1871 there remained only a little more than two and a half million acres, and by 1883 only a million. In that year a well-publicized incident demonstrated the inadequacy of the existing laws to protect the public interest. At a sale of state lands held in Madison a group of speculators who, according to the theory of the law, should have been bidding against one another to raise the price to a fair level, formed a pool and appointed a committee to bid for all of them. The committee obtained a large block of pineland at the minimum price of three dollars an acre and later presided over a private auction at which the members of the pool actually did bid against one another. Some of the land the state had just sold for the legal minimum brought fifty dollars

Referred So Much of the Governor's Message As Relates to Swamp Lands (Madison, 1855), 7, 10; *La Crosse Independent Republican,* August 16, 1854, January 10, February 21, 1855; *Compilation ... Relating to the Swamp and Overflowed Land Fund,* 210.

[14] Mitchel and Osborn, *Geographical and Statistical History of the County of Winnebago,* 117; *La Crosse Independent Republican,* March 23, 1859; B. Cornell to Ingram and Kennedy, October 4, 1856, in the Ingram Papers; *Report of the Commissioners of School and University Lands,* 1865, pp. 15–16; 1876, p. 4; 1878, p. 13; *General Laws,* 1867, ch. 50, 51, 139; 1869, ch. 77, 142; 1873, ch. 75; *Governor's Message,* 1871, p. 6; *Eau Claire Free Press,* January 19, 1871, January 18, March 29, September 13, 1883; *Compilation ... Relating to the Swamp and Overflowed Land Fund,* 73, 236.

WHi(X3)6020

Vicinity of Marshfield, Wisconsin, about 1888. Upham Lumber Co. logging train and crew, including engineer Frank Lueckenback. The "Old Vanderbilt" engine shown was formerly the Wisconsin Central RR engine #47; it was originally the NYP&B RR engine "Oregon," renamed "C. Vanderbilt," built in Taunton, Massachusetts in 1847.

James L. Gates

Courtesy State Historical Society of Wisconsin

Cartoon of James L. Gates, from *Wisconsin Men Who Do Things* (undated), illustrating the positive attitude toward the clearing of Wisconsin's forests held by many people at the time the logging was taking place.

at the private sale. When the state land commissioners learned of the conspiracy, they cancelled the sale and thereby brought down upon themselves the denunciation of the newspapers of the logging counties. This means of dealing with fraud was denied the commissioners after 1884, when the state Supreme Court ruled that they might not withdraw from the market any lands that had once been offered for sale.[15]

But even this policy of rapid disposal did not content prospective buyers. They tried repeatedly to prevail upon the legislature to transfer portions of the state lands to the control of county governments. A bill ordering the outright donation of all swamplands to the counties in which they were situated passed the legislature in 1862, but was vetoed by Governor Louis P. Harvey on the ground that in counties where logging was the chief activity the timbered swamplands would be at the disposal of the lumbermen, who in all such counties held absolute control of the government. Later governors were less concerned. Several counties within the white pine area acquired their drainage lands by legislative act before 1890 and eventually disposed of them on terms even more advantageous to the buyer than those of the state.[16]

The federal government's grants of large areas of white pine land to promote internal improvements and the rapidity with which speculators acquired the rest of the public timberlands forced the average logger to buy much of his stumpage from railroads and other private owners. One of the largest holdings of forest land in Wisconsin was that of Cornell University. Under the Morrill Act of 1862 the states that no longer had any public lands were given scrip which could be used to locate

[15] *Chippewa Herald,* February 25, 1871, September 21, 1883, February 27, 1885; *Eau Claire Free Press,* January 18, March 22, 29, September 13, November 29, 1883, April 17, 1884; Oshkosh *Northwestern,* September 20, November 8, 1883; State *ex rel.* Holston *v.* Commissioners of Public Lands, 61 Wis. 274 (1884). The figures on unsold lands do not include lands sold by the state but not yet paid for.

[16] Copy of Bill 62S, in the John T. Kingston Papers; *Special Message of the Governor Returning with Objections Bill Number 62, Senate* (Madison, 1862), 6; *General Laws,* 1874, ch. 322; 1880, ch. 261; *Chippewa Falls Democrat,* February 2, 1871; *Chippewa Herald,* February 11, 1871, December 21, 1888; *Eau Claire Free Press,* March 27, 1890; Stephenson, *Recollections,* 191–192.

lands in other states. Most of the states sold their scrip at once for what it would bring on the market, usually about thirty to sixty cents an acre, but New York pursued a different course. With its thirty-three members of Congress it was entitled to almost a million acres of the public domain. The state was induced by Ezra Cornell, a wealthy businessman, to let him locate lands in Wisconsin and other Western states so that the educational institution at Ithaca in which he was interested might benefit from their increase in value. About a half million acres were selected in Wisconsin after expert landlookers had reported on them.[17]

Cornell University maintained agents in Wisconsin to manage its lands and sell them as it became desirable to do so. By 1880 the large lumber companies were beginning to make extensive purchases of the Cornell holdings. In that year Knapp, Stout and Company paid an average of fifteen dollars an acre for some thirty thousand acres. Two years later the Chippewa Logging Company paid nearly two million dollars for about a hundred thousand acres. During the biennium ending in January, 1883, the university sold 170,939 acres for $2,644,565, and the stumpage on 14,638 acres for $98,605. By 1922 Cornell had received over seven million dollars from its Wisconsin lands, more than five million of which was clear profit.[18]

As the public timberlands in Wisconsin began to dwindle the large lumber companies became more and more preoccupied with land transactions. By 1890 the lumbering firms that had engaged in land speculation were beginning to withdraw large tracts of timber from the market in anticipation of the time when the pine would be exhausted. The amount of unappropriated federal land in Wisconsin steadily declined. In 1889 the government abolished its land offices at La Crosse and St. Croix Falls for lack of business. In 1896 there remained in Wisconsin

[17] *Eau Claire Free Press,* February 1, 1883; *Report of the State Forester,* 1906, p. 53; Gates, *Wisconsin Pine Lands,* 1–106.

[18] *Report of the State Forester,* 1906, p. 53; *Chippewa Herald,* October 5, 1872, May 24, December 5, 1873, September 22, 1882, October 7, 1892; *Eau Claire Free Press,* November 16, 1876, December 6, 1877; *Eau Claire Leader,* January 29, February 5, 12, 1922; Gates, *Wisconsin Pine Lands,* 208–243.

only a half million acres of unappropriated federal land, in 1899 three hundred thousand acres, and in 1914 eight thousand.[19]

The land policy of the federal government encouraged a great squandering of timber resources even in excess of what was the inevitable result of the frontier psychology. Until almost 1900 standing timber on government land was so cheap in Wisconsin that many lumbermen were content to cut trees in advance of any real economic demand, to utilize only partially the trees they had felled, and to leave inferior species in the forests to be destroyed by the slash fires that usually followed the loggers. Unwise local taxation encouraged these wasteful practices. Moreover, there was no incentive to reforest cutover lands. Under the land system of the nineteenth century private enterprise could not afford to make the long-term investment required to grow new forests and make a profit from them. Under a system of licensed cutting the government could have charged the lumbermen a reasonable price for their logs and at the same time have maintained permanent forests.

Not only did the low price of public timberlands, as compared with that of similar lands privately held, encourage speculation, but their steady increase in value after being sold by the government promoted monopoly. Throughout the nineteenth century and the early part of the twentieth the price of standing timber on the open market increased tremendously. Whereas in 1855 Wisconsin pineland sold on the open market for an average of less than five dollars an acre, by 1913 the price exceeded twenty dollars, and some choice acreage brought as much as fifty. A tract of Wisconsin pineland bought in 1894 for the government price of a dollar and a quarter an acre was sold in 1908 for thirty dollars an acre. It was not uncommon for the price of a timber tract to double within two years. Many of the lumber fortunes

[19] *Report of the Commissioner of the General Land Office,* 1896, pp. 187–188; 1899, p. 214; 1903, p. 5; 1914, p. 97; [Wisconsin] Committee on Land Use and Forestry, *Forest Land Use in Wisconsin,* 14; *Eau Claire Free Press,* July 29, 1886, June 13, 1895; La Crosse *Weekly Chronicle,* May 22, 1890; *Northwestern Lumberman,* March 2, 1889. Preoccupation with land-buying is revealed in numerous letters of the Holt Lumber Company written after 1880. Holt Lumber Company Papers.

represent the profits not from a manufacturing enterprise but from the resale of land bought at a low price.[20]

Unbridled speculation tended to build up great timberland holdings that ultimately resulted in an unhealthy concentration of ownership. Such concentration was furthered by federal land grants in that it placed large blocks of land in the hands of private corporations, particularly railroads. Thousands of acres were purchased by such magnates as Isaac Stephenson, Philetus Sawyer, and Frederick Weyerhaeuser; Eastern speculators held many thousands more. In 1875 the Menominee River Lumber Company, of which Sawyer was the chief stockholder, owned seventy-five thousand acres of pine, and this was only one of his many interests. The companies controlled by Weyerhaeuser owned even more land. As early as 1869 the greater part of the pine forest in the Wolf River district was owned and controlled by a few millowners. In 1913, near the close of the lumbering era, ten owners held twenty-four per cent of all the standing timber in Wisconsin, and ninety-six owners held three-fourths of it. Most of these men also held large tracts in other forested regions of the country.[21]

A few strong interests controlling the greater part of the standing timber in an area could set the price of logs and their products there. Landholding, not large-scale manufacture, was the key to monopoly in the lumber industry. Large holdings enabled their owners to control not only the small holdings interspersed among them but every sawmill to which they were tributary. Interlocking directorates, ownership of subsidiary companies, and close affiliation with other lines of business augmented the power of the large timberland owners.

The ownership of timberlands was, however, complicated by vexatious costs. One of these was the interest charge on capital

[20] Bureau of Corporations, *The Lumber Industry,* pt. 1, pp. 1, 26, 38–39; Ise, *United States Forest Policy,* 359, 361.

[21] Oshkosh *Northwestern,* June 17, 1875; *Oshkosh Journal,* November 6, 1869; *Eau Claire Free Press,* March 17, 1881; Bureau of Corporations, *The Lumber Industry,* pt. 1, pp. xxi, 22, 182; Hugh Bellas to Orrin H. Ingram, September 11, 1896, in the Ingram Papers; Ise, *United States Forest Policy,* 329–332, 355; Paul W. Gates, "Homestead Law in an Incongruous Land System," in the *American Historical Review,* 41:666 (July, 1936).

borrowed for speculative purposes. Another was the expense of fire protection, although the annual cost rarely exceeded a tenth or a fifth of a cent per thousand feet of stumpage. Local taxes were sometimes high. Taken together, these costs rapidly increased the investment of a speculator in timberland. By the turn of the century, however, the rising value of the land had absorbed these expenses and paid a generous profit.

His taxes were perhaps the expense that was most annoying to the timberland owner, especially if he did not control the election of the assessors. Many small communities made it a practice to levy heavy taxes on the timber of absentee owners. Some regarded the forests as a ready source of revenue for various projects which the speculator regarded as sheer extravagance. Schools, roads, and bridges infinitely superior to those in comparable communities elsewhere graced many an agricultural settlement fortunate enough to be situated within a forest tract. Some forest land was taxed excessively either because the assessor was ignorant or because he was dishonest enough to use the threat of still higher taxes to blackmail landowners. Unscrupulous adventurers were attracted to some counties, where they "crowded" themselves into town and school offices and began a "system of thievery." Other inequalities of assessment were no less exasperating. Great areas of land within the railroad grants were tax-exempt for many years, with the result that the railroad companies were in a position to realize greater profits than the private speculators from the sale of their timber.[22]

Since the timberland speculator did not expect to benefit from most of the improvements made to attract settlers, the propensity of some counties to enter heavily into debt was also frowned upon by the speculator. In 1889, when the population of Ashland County was only about eight hundred, it had a debt

[22] *Resources, Advantages and Productions of the Counties of Brown, Door, Oconto and Shawano, in the State of Wisconsin* (Green Bay, 1870), prepared and published by order of the county boards of supervisors, 31; *Hand Book of Price County,* prepared by the state Board of Immigration under the direction of the county board (Milwaukee, 1896), 17; Bruncken, *Forests and Forestry,* 210–213; Ise, *United States Forest Policy,* 357; *Legislative and Congressional Acts Concerning Land Grants,* 26–27.

of two hundred thousand dollars; and Lincoln County, with a population of less than a thousand, carried a debt of a hundred and ten thousand dollars. In 1892 the little town of Eagle River, whose population barely exceeded a thousand, levied taxes in excess of fifty thousand dollars.[23]

Landowners resorted to many devices to defend themselves against high property taxes. They established numerous associations for contesting them. They made efforts to have counties divided in the hope of getting control of the assessing machinery in the new units. They fought the incorporation of urban units whenever it offered the prospect of higher taxation. In 1898 lumbermen contributed from their own pockets toward the work of a state tax commission which was collecting data on local taxing methods and which, needless to say, was friendly to lumbermen and landowners. And when all else failed, many timberland owners stripped their lands of the trees and allowed the worthless stumpland to revert to the county. This practice not only rendered thousands of acres prematurely worthless but destroyed a vast amount of timber long before there was any economic demand for it.[24]

Despite all the complaints, however, it is probable that the lumberman benefited from underassessment more often than he suffered from overassessment. As Governor Harvey recognized, it was not difficult for a group of lumbermen to control the politics of a county. As early as 1855 the assessed valuation of land in some lumbering counties was less than half its equalized valuation. In 1868 the secretary of state complained that in ninety per cent of the counties of Wisconsin the assessment returns gave evidence of "palpable violations of the law" where the offenders realized that they were fairly safe from apprehension. A great deal of property was omitted altogether from the

[23] *Eau Claire Free Press,* February 13, 1879; La Crosse *Weekly Chronicle,* February 25, 1892; *Northwestern Lumberman,* September 15, 1877.

[24] T. B. Goodrich to Holt and Balcom, March 2, 1887, and Augustus Cole to Holt and Balcom, March 8, 1887, in the Holt Lumber Company Papers; Charles F. Bone to C. A. Chamberlin, March 4, 1885, and K. K. Kennan to Orrin H. Ingram, January 24, 1898, in the Ingram Papers; *Chippewa Herald,* April 8, 1871, June 12, 1874, April 24, 1891; La Crosse *Weekly Chronicle,* December 8, 1887; *Eau Claire Free Press,* May 17, 1900.

tax rolls, he said, and much of what was taxed had been assessed at only half its true value. In 1891, the same year that an Oshkosh newspaper was openly advocating that manufacturing interests be put in control of the municipal government to insure their greater satisfaction with taxes, some of Wisconsin's pinelands were being assessed at less than a fifth of their market value. When asked whether he would introduce a bill to put a stop to such unfair assessment, a legislator replied, "I can't do it. I have several local measures that I dare not jeopardize. They are not of as great importance to the commonwealth as this but my constituents would not see it as do you and I." The town of Baxter in Wood County was for years under the control of a single lumberman who assessed his own property low and that of outsiders high. Many early lumbermen tried to discourage settlement in the neighborhood of their lands by issuing information that the region was unsuited to agriculture.[25]

As a result of lumbermen's practice of cutting the timber from their lands and then failing to pay their taxes, Wisconsin counties held more than a quarter of a million acres on tax deeds by 1898, and "considerably more" was encumbered with outstanding tax certificates, the redemption of which seemed very unlikely. The State Forestry Commission recommended that the counties retain the title to the denuded lands and replant them with trees rather than sell them "at any price whatsoever." But many of the counties in which large areas yielded no revenue were too crippled financially to do even this.[26]

Unfit for agriculture as were the sandy "pine barrens" of northern Wisconsin, many counties nevertheless tried to sell their tax lands to foreign immigrants. At the end of the eighties lumber companies, realizing that the agricultural immigrant was a potential buyer of otherwise worthless cutover lands, began selling them to farmers instead of letting them revert to the

[25] *Governor's Message,* 1868, p. 4, quoting the secretary of state; *Eau Claire Free Press,* March 5, 1891; *Report of the Secretary of State,* 1855, pp. 60–61; Oshkosh *Weekly Northwestern,* January 1, 1891; *Chippewa Herald,* March 28, May 9, 1884; *Wisconsin and Its Opportunities* (n.p., 1911), a pamphlet issued by the Soo Line, 7; Schafer, *Winnebago-Horicon Basin,* 294–295.

[26] *Report of the Forestry Commission of the State of Wisconsin* (Madison, 1898), 19; *Chippewa Herald,* March 20, 1874.

counties. J. L. Gates, a Milwaukee lumber dealer, made the "conversion of these lands a life business." By 1899 he owned nearly three quarters of a million acres of cutover lands and was still buying them. Newspapers and pamphlets, using the publicity methods of the day, sought to eradicate the idea that crops could not be grown on denuded pinelands. The state, undoubtedly acting in good faith, aided the movement by creating an immigration bureau, which attracted many foreign immigrants into northern Wisconsin. Unscrupulous promoters sold worthless lands to foreigners and to people who had no experience in farming. After a few years of struggle most of them abandoned their farms or resigned themselves to hopeless poverty. These conditions and the consequent demoralization of the settlers threatened to create in Wisconsin a class of crackers that was to become a burden, financially and socially, upon the commonwealth.[27]

[27] Bruncken, *Forests and Forestry,* 190–192; James L. Gates to Orrin H. Ingram, July 17, 1899, in the Ingram Papers; *Eau Claire Free Press,* February 1, September 20, 1885, November 27, 1890, June 20, 1895, February 10, 1898, March 8, 1900; *Milwaukee Sentinel,* May 30, 1899; Milwaukee *Evening Wisconsin,* May 31, 1899. For a recent study of one group of immigrants who settled on the cutover lands, see John I. Kolehmainen and George W. Hill, *Haven in the Woods: The Story of the Finns in Wisconsin* (State Historical Society of Wisconsin, Madison, 1951).

11. Failure of the Land Laws

THE incompatibility of the land-sale laws with the economic realities of the lumber industry was a strong incentive to commit fraud, which it is hard not to regard with leniency. The simple truth is that under ordinary circumstances the acquisition of timber holdings under either the pre-emption or the homestead law was a crime. Both the pre-emptor and the homesteader had to take an oath that the land he was acquiring would be used for his own benefit and that he had "improved" it, presumably by agricultural cultivation. Yet thousands of acres of sandy land in northern Wisconsin were as ill adapted to agriculture as they were well suited to the growth of excellent pine timber. Almost any homestead or pre-emption entry on this land during the last half of the nineteenth century may be taken as *prima facie* evidence of fraudulent intent, for no competent farmer in his right mind would have considered this soil for agriculture when large fertile areas elsewhere on the frontier were still unsettled. As the commissioner of the General Land Office put it in 1877: "The homestead and pre-emption laws are now educating thousands of men in the crime of perjury. It would be better to pass a law granting the land [free] to persons who would manufacture the timber upon it into lumber . . . as that is in fact what they do and all they do now after taking them under the homestead and pre-emption laws."[1]

[1] *Report of the Commissioner of the General Land Office*, 1877, p. 35. A soil

It may well be that the homestead and pre-emption laws would have been regarded favorably by some lumbermen even if the direct-sale law had been better adapted than it was to logging practices, for both acts offered decided advantages to timber buyers unhampered by too many scruples. Either law offered a prospective buyer prepared to violate its terms the opportunity to acquire a desirable tract of timberland without running much risk that a competitor would discover his intention and anticipate him at the land office. Moreover, he could get the timber on it without paying even the minimum price stipulated by the direct-sale law.

The pre-emption law allowed the individual farmer, in whose interests it had been adopted, to stake out a claim on the public domain before the land came up for auction under the law of 1820. By employing such a "farmer" as its dummy a lumber company could obtain the staked claim before competitors could get it. A pre-emption claim could of course be purchased also from a farmer who was not a hireling but a legal entrant for the land; such a transaction, too, obviated the necessity of facing competitors at a public sale. The homestead law, which allowed the homesteader to purchase his entry at the usual minimum price after living on it for six months, could be used in exactly the same way.[2]

Even more advantageous for the timber buyer was the provision of the pre-emption law which extended credit to the applicant for thirty-three months after he had made his "declaration of intention." Concretely, this meant that for this length of time the land in question would not be offered at public sale because a citizen had recorded his intention of "proving up" a pre-emption claim; he need have paid no money to the government. During the waiting period the pre-emptor could strip his

survey of Wisconsin forest reserve lands made in 1910 indicated that only twenty-two per cent of the land could be cultivated profitably. *Report of the State Forester,* 1909–11, p. 56.

[2] The commutation clause permitted the settler, if he wished, to treat his homestead as a pre-emption claim after six months' residence (fourteen months after 1891). The purpose was to protect the interests of homesteaders when changed circumstances made it impossible for them to remain on their claims for five years. *United States Statutes at Large,* 12:393; Bureau of Corporations,

claim of all valuable timber without incurring any serious risk of being prosecuted as a trespasser. Once he had removed the timber, the declarant merely abandoned his claim without bothering to complete his transaction with the government. A lumber company using dummies in this way could obtain timber for the cost of the necessary labor. Nor was it always necessary to use dummies; independent pre-emption claimants could strip their claims just as easily and sell the timber to a lumber company at a bargain price.[3]

Perpetrators of such frauds used incredibly transparent devices to remain within the letter of the law. To meet the provision of the homestead law that an entryman must have reached the age of twenty-one unless he was the head of a family, a lumberman using an unmarried minor as a dummy would take the boy into the land office, mark "21" on the floor with chalk, stand him upon the number, and have him swear (truthfully!) that he was "over twenty-one." Both the pre-emption and the homestead law also required that the claimant produce satisfactory evidence of habitation and improvement at the time of entry. To meet this stipulation the pseudo-agrarians constructed any kind of crude shack that could be described as a house. A not uncommon practice was to set up a large dry-goods box, cut little holes in the sides for "windows," and place a few shingles on top for a roof. This enabled the applicant to swear that he had a "good board house, 14 by 16, with a shingled roof, doors, windows, etc."; he did not explain, of course, that the "14 by 16" represented inches, not feet. By placing a whiskey bottle in each "window" the applicant added a touch of luxury; he could then swear that the house had glass in its windows. Baffled by such artifices, a land-office inspector complained in 1885 that "there should be some way to distinguish between a fire guard of a few furrows plowed around a quarter section and a cornfield—some way to determine whether a description of a house

The Lumber Industry, pt. 1, pp. 259–260; Benjamin H. Hibbard, *History of Public Land Policies* (New York, 1924), 389.

[3] *United States Statutes at Large,* 16:279; *Report of the Commissioner of the General Land Office,* 1875, p. 10; 1882, p. 207; Bruncken, *Forests and Forestry,* 191; *Chippewa Herald,* March 29, 1873.

'14 x 16' referred to inches or feet square; whether the floor was *bored* or *board,* and whether the 'shingle roof' meant more than two shingles, one on each side."[4]

Such barefaced frauds could not have succeeded, of course, without the connivance of the land-office officials. William Sparks, who became commissioner of the General Land Office in 1885, was shocked by the discoveries he made when he took office. "The vast machinery of the land department," he wrote, "appears to have been devoted to the chief result of conveying the title of the United States to public lands upon fraudulent entries under strained constructions of imperfect land laws and upon illegal claims under public and private grants." A year later Elihu B. Washburne, former congressman of Illinois and "watch-dog of the treasury," described the land office as the "most corrupt department that ever existed in any government on the face of the earth." Commissioner Sparks admitted frankly that the land-office agents in many districts had no morals where the land laws were concerned, but there is little evidence that any of them were punished for connivance in fraud. In 1872, for example, the land office at Eau Claire was closed for several months while an inspector from Washington investigated charges that its register and receiver were using their powers to aid a select "pineland ring." The local officials were exonerated, the only discipline administered them being the admonition to give greater attention to their duties and allow a certain clerk less freedom in the office.[5]

Exactly what proportion of homestead and pre-emption entries in Wisconsin were fraudulent is impossible to estimate from the records. The head of the General Land Office admitted in 1875 that the perpetration of frauds through the misapplication of the laws had become a tradition in his bureau. The next

[4] Nelligan, *Life of a Lumberman,* 175–176; Bureau of Corporations, *The Lumber Industry,* pt. 1, p. 261; *Report of the Commissioner of the General Land Office,* 1885, p. 52.

[5] *Report of the Commissioner of the General Land Office,* 1885, pp. 3–4, 51; La Crosse *Weekly Chronicle,* February 4, 1886; *Eau Claire Free Press,* January 4, 11, 25, April 4, May 2, 1872, April 5, 26, 1877; La Crosse *Republican and Leader,* January 6, 13, 27, April 27, 1872; *Chippewa Herald,* January 6, 27, 1872, August 21, 1874, March 23, April 13, 20, 1877; Warren, *Pioneer Woodsman,* 44–46.

year he reported that "in all the pine region of Lake Superior and the Upper Mississippi, where vast areas have been settled under the pretense of agriculture under the homestead and pre-emption laws, scarcely a vestige of agriculture appears."

During 1879 ninety per cent of the homestead entries in the white pine area of the Great Lakes, the commissioner reported, were made for the purpose of "stripping the land of its timber," and fraud had become so prevalent that honest lumbermen could hardly compete with those whose consciences were less squeamish. That no more than twenty to twenty-five per cent of the homestead and pre-emption entries in the Wausau and Eau Claire districts in the years 1885 to 1888 had been made for the cutting of timber was attributed by local officials to the vigorous campaign the General Land Office had been conducting to end such abuse of the laws.

The campaign certainly produced no permanent abatement of fraud. In 1888 ninety-nine of every hundred pre-emptions in the Wausau land district, and ninety in the Eau Claire district, were fraudulent. In 1890 twenty-six cases of fraud were pending in the federal district courts of Wisconsin, and it would not be unreasonable to assume that many more violations of the law escaped the attention of the courts. Twenty years later, when the lumber industry of the state was nearing its end, the chief of field service in the General Land Office declared that he had personally examined "solid townships" in northern Wisconsin in which virtually all the even-numbered sections had been acquired under the homestead law and stripped of their timber, and yet "not a single voter or inhabitant could be found in the township."[6]

Appalled by such widespread and palpable fraud, the various commissioners of the General Land Office, from about 1870 on, repeatedly urged Congress to pass remedial legislation. Their appeals produced no result until 1877, when funds were finally appropriated for curbing the abuses. The vigor with which Sec-

[6] *Report of the Commissioner of the General Land Office,* 1875, p. 10; 1876, p. 7; 1879, p. 186; 1888, pp. 78, 87; 1890, p. 340; Bureau of Corporations, *The Lumber Industry,* pt. 1, p. 261.

retary of Interior Carl Schurz put the appropriation to use apparently aroused legitimate lumbermen from the indifference with which they had hitherto regarded violations of the law. In any event they joined the conservationists during the eighties and nineties in obtaining from Congress a gradual increase in appropriations for eliminating fraud. At the same time the General Land Office formulated more stringent regulations for homesteaders. Realizing at last that all such measures were inadequate to remedy the situation, Congress finally, in 1889, suspended cash entries altogether in every public-land state except Missouri. Two years later the pre-emption and direct-sale laws were repealed, and the Homestead Act was modified to make the perpetration of fraud more difficult.[7]

Technically the fraudulent entry of government land under the pre-emption and homestead laws for the purpose of cutting the timber on it could be classified as timber trespass. The federal law defined as trespass any illegal destruction or removal of timber on the public lands or any act encouraging others to commit these crimes. The outright theft of government timber without the pretense of acquiring title to the land was an even more serious menace to the forest resources of the country than fraudulent land entries. In the face of the frontier psychology —the firm conviction that the public lands belonged to the people as much as "the sunshine and the air"—such removal of trees from the public domain was inevitable.[8]

The federal government was slow to develop a system for the suppression of trespass. During the colonial period the British policy of reserving the best trunks for the royal navy had been so unpopular in the pine forests of New England that it became one of the many causes of the American Revolution. Accordingly the new government of the seaboard states was not dis-

[7] Ise, *United States Forest Policy,* 100; *Chippewa Herald,* August 11, 1882, September 11, 1885; *United States Statutes at Large,* 25:854; 26:1095.

[8] Thomas Donaldson, *The Public Domain: Its History with Statistics* (47 Congress, 2 session, House Miscellaneous Document no. 45, pt. 4, serial 2158, Washington, 1884), 1060–1061; Bruncken, *Forests and Forestry,* 193; Warren, *Pioneer Woodsman,* 56–57; Milo M. Quaife, ed., *An English Settler in Pioneer Wisconsin: The Letters of Edwin Bottomley, 1842–1850 (Wisconsin Historical Collections,* vol. 25, Madison, 1918), 68, 73.

posed to adopt any policy of forest regulation. Not until 1807 was prohibition of trespass entered on the federal statute books, and then only as a part of another law designed to prevent settlement in advance of survey. The first act designed specifically for the protection of the trees themselves was one passed in 1817 to insure a reserve of live oak and red cedar for the use of the navy. In 1821 Attorney General William Wirt held that the statute of 1807 sanctioned the use of the army to protect federal timber, whereupon the General Land Office instructed its local officials to arrest trespassers. In 1831 Congress passed a law forbidding the unauthorized removal of any timber from the naval reserve lands, and, if the enacting clause was broadly interpreted, from any other lands owned by the United States. Under this statute the solicitor of the treasury, a newly created official, carried on an active campaign against trespassers. In 1850 the Supreme Court, to which the case of a Michigan trespasser had been appealed, held that the statute applied definitely to the entire public domain of the United States and not only to reserve lands of various kinds.[9]

In 1849 the Home Department (later the Department of the Interior) was created and charged with the supervision of all public lands. The next year the new secretary, influenced by the recent Supreme Court decision, appointed the first special agents to deal with trespass. Because at that time most of the trespassing occurred in the pine forests of the Great Lakes region, the new agents were assigned only to Wisconsin, Michigan, and Iowa. Wisconsin at first had one agent, Iowa one, and Michigan two, but more agents were soon appointed. Thus by the mid-nineteenth century the United States had a fairly well-developed policy for the control of trespass.[10]

The new agents found, however, that they faced almost insuperable difficulties, for by 1850 trespassing had become an

[9] Donaldson, *Public Domain,* 357; United States *v.* Briggs, 9 Howard 351 (1850); Cameron, *Governmental Forest Control,* 125–129; *Report of the Commissioner of the General Land Office,* 1866, p. 33; 1874, p. 40; *United States Statutes at Large,* 2:445; 4:472.

[10] Cameron, *Governmental Forest Control,* 130–132; *Report of the Commissioner of the General Land Office,* 1874, p. 40; 1877, p. 16.

accepted practice in the Great Lakes pinery. More than a decade earlier Moses M. Strong, who had become federal attorney for the Territory of Wisconsin in 1838, found letters to his predecessor informing him that "a large number of men were engaged in cutting and carrying off the valuable pine Timber from a portion of the public domain within his district." Strong, who had been ordered by the solicitor of the treasury to suppress trespass, hardly knew what was meant by his instructions to regard more favorably the people who were "actually inhabiting and cultivating the public lands," a concession born of the series of pre-emption laws passed after 1830. The trespassing lumbermen were "actually inhabiting" the public lands, and most of them grew a few vegetables and grains to feed their men and animals. Were they therefore to be regarded as pre-emptors?[11]

His question was not definitely answered until 1845, when, after the passage of the general pre-emption law, the attorney general held that pre-emptors might cut from their claims such timber as was needed to clear their fields and meet their domestic requirements; presumably they could cut no more until they held title. Meanwhile Strong instructed the United States marshals in Wisconsin to enter complaints against "all persons found cutting timber on Government Land," but he failed utterly to stop trespassing. In the lead district of southwestern Wisconsin, where trespassers were numerous in the 1840's, the government's measures were met with "a general spirit of resistance to the execution of the laws." In 1844 a correspondent from Galena wrote to a congressman urging that the pinelands on the St. Croix and Chippewa be brought into the market "as soon as practicable," for otherwise the "outrageous waste being committed on them" would soon destroy their value. At the same time privately owned timberlands were also being ravaged by extensive trespassing. Previous to 1850, at least, the government's efforts to eradicate the evil were a failure.[12]

[11] Moses M. Strong to H. D. Gilpin, November 20, 1838, in the Strong Lumber Papers.

[12] Benjamin Piper to Moses M. Strong, January 13, 1840, and A. Graham to Strong, January 3, 1840, in the Strong Lumber Papers; *Report from the Secretary of the Treasury Showing . . . the Number of Suits Which Have Been Insti-*

This failure can be attributed largely to the frontiersman's attitude toward land. No Westerner regarded trespass as a serious crime. To take trees from the public domain was no more immoral than it was to float a canoe on a public river. Agricultural settlers and lumbermen of undoubted integrity had no scruples whatever about "visiting Uncle Sam" to get some wood. Public opinion in Wisconsin, especially before the Civil War, condoned all forms of trespass and upheld the trespasser, even the lumbermen who drew their entire stock of logs from government lands and those who entered a few forties in order to "make a show" of owning timberland while they took most of their logs from adjacent public lands. One private owner who sought to bring criminal action against a trespasser was advised by a lawyer, "Now don't try that. All of these fellows have had 'some of them hams,' and you can't get a jury in all that country to bring you in a verdict of guilty, no matter how great and strong the evidence." In 1843 an English settler, commenting on the matter of trespass in a letter home, reported that "we are not afraid of Informers here for it is a general Practice . . . followed By both natives and new settlers." The pioneer lumberman was often not even aware that he was a trespasser. In any event he was, to his own way of thinking, "earning by his strong right arm" his title to the trees.[13]

And so when in 1850 the government began a more vigorous campaign against trespass an explosion was inevitable. Early in November the federal marshals began seizing logs and lumber that had been cut illegally from government lands in Wisconsin. The public reaction was immediate and violent. Newspapers stormed that the hardy lumberman, who was cutting timber with the "tacit sanction" of the previous administration, was being made the "prey of rapacity and avarice." "Spies" were being sent to inform on him, his lumber was being confiscated

tuted against Citizens of Illinois and Wisconsin Since March 4, 1841, for Trespasses upon the Public Lands (28 Congress, 1 session, Senate Document no. 37, serial 432, Washington, 1844); *Trade and Commerce on the Upper Mississippi* (28 Congress, 1 session, Senate Document no. 242, serial 434); *Report of the Commissioner of the General Land Office,* 1866, p. 33.

[13] Warren, *Pioneer Woodsman,* 58; Quaife, *English Settler,* 40–41; Folsom, *Fifty Years in the Northwest,* 703.

without warrant or legal process, and he was being forced to buy back his own property at extortionate rates. The harried lumbermen sent an agent to Washington to protest against this "monstrous exaction," and the Western representatives in Congress swung into action.[14]

On the twenty-second of July, 1852, Benjamin C. Eastman of Wisconsin rose from his seat during the House debate on the rivers and harbors bill to laud the accomplishments of the frontier lumbermen. They had, he said, made their part of the West more flourishing than some richer parts of the country by their courageous, unceasing labors, and the government ought,

> for the present, certainly, to leave this enterprising class of citizens to use a portion of the timber of the public lands without stint, as a sort of bounty for the hardship they have undergone in piloting the way to this country. But very far from this is the fact; for the whole power of the country, in the shape of the United States marshals and a whole *posse* of deputies and timber agents, appointed by the President, without the least authority of law, have been let loose upon this devoted class of our citizens, and they have been harassed almost beyond endurance with pretended seizures and suits, prosecutions and indictments, until they have been driven almost to the desperation of an open revolt against their persecutors.

Three months later Delegate Henry H. Sibley of Minnesota Territory inveighed against the "outrages in Wisconsin and Minnesota" and the laws under which they were being perpetrated, laws which he should like to see "expunged from the statute books as a disgrace to the country and to the nineteenth century." "Sir," he thundered, "the time is at hand when the arbitrary exercise of power, such as I have alluded to, will be rebuked." And Senator Augustus Caesar Dodge of Iowa, who had voted for the creation of the Department of the Interior in 1849, now expressed his deep regret that he should have helped "to extend the power of the government."[15]

[14] *Fond du Lac Journal,* November 28, 1850; *Wisconsin Argus* (Madison), July 9, 1851; *Wisconsin Express* (Madison), July 10, 1851; Thomas J. Moorman to Moses M. Strong, July 19, 1852, in the Strong Lumber Papers; Usher, *Wisconsin,* 1:209.

[15] *Congressional Globe,* 32 Congress, 1 session (1851–52), Appendix, 486, 851; Cameron, *Governmental Forest Control,* 139.

When Secretary of Interior Alexander Stuart realized what a hornet's nest he had stirred up, his zeal for reform subsided and he transferred to his subordinate, the commissioner of the General Land Office, the immediate responsibility for protecting the public domain. Early in 1854 this official issued instructions to his timber agents to seize all logs, floating timber, and lumber that had been stolen from government lands and sell them at public auction. Bona fide settlers might continue unhindered to clear land for the establishment of homes, but with trespassers there was to be no compromise. No trespasser was to be allowed to keep stolen timber simply by paying for the logs; if he did not wish to forfeit them, he must also buy the land from which they had been taken. And no longer might anyone obtain permission to cut trees on public land by paying a fee in advance. Where legal process existed, seizure of the timber was to be made by the United States marshals, and criminal suit was to be brought by the federal district attorneys.[16]

The lumbermen, having thus far dealt with timber agents who had "compounded and compromised" with them, were greatly exercised over the new instructions. They were in no mood to submit peacefully to what they considered an infraction of their rights. In eastern Wisconsin no one would bid at a public sale of seized timber without the consent of the man from whom it had been taken. In May of 1854 a deputy agent was shot dead near Dubuque as he was trying to take possession of stolen logs on the Mississippi. A "systematic warfare" was begun. Daily the Chicago newspapers attacked the agents in Wisconsin, whose conduct they described as oppressive in the extreme. The *Chicago Tribune* even advised resistance by force. The lumbermen held meetings at which "violent harangues" were delivered, following which they resorted to violence and force to prevent seizures and arrests. An agent in western Michigan had to call upon bluejackets from a naval vessel for aid in making arrests.[17]

[16] *Depredations on Timber Lands* (33 Congress, 1 session, House Executive Document no. 115, serial 727, Washington, 1854), 4–6.

[17] Madison *Daily Argus and Democrat*, May 16, 1854; *Depredations on Timber Lands*, 9.

Early in 1854 Eastman introduced a resolution calling upon the secretary of the interior for complete information on the methods being pursued to prevent trespass in the Great Lakes region. The secretary responded with a full report. Admitting that he had not been definitely authorized to appoint timber agents, he argued that they had been essential to his enforcement of the law of 1831 as interpreted by the Supreme Court. His report included accounts written by the agents whose territory lay in Wisconsin. Sixteen sawmills on the Black River had "until within the last year" been supported by logs taken from the public lands; together they had consumed some two hundred million board feet of government-owned pine. Eight mills on the Chippewa and Red Cedar rivers were cutting annually about two million board feet from the public domain, and many more logs were floated down the Mississippi for sawing. The same state of affairs existed on the St. Croix. Of the five hundred million feet shipped from eastern Wisconsin and northern Michigan in the preceding decade, almost ninety per cent, it was estimated, had been stolen. Twenty-one indictments had been returned against trespassers in Wisconsin and one against an earlier agent who had illegally sold the right to cut timber on government land. So common had the practice of selling "stumpage privileges" been before 1854 that most lumbermen thought they could buy standing timber from the government without having to buy the land, and most of them were very well satisfied with this arrangement. After the government's enforcement campaign had begun, however, "large bodies" of land had been entered in both Wisconsin and Minnesota.[18]

The storm loosed in the West by the actions of the timber agents in the Great Lakes region was politically embarrassing to the Democratic Pierce administration. In the interest of peace the Whig commissioner of the General Land Office, John Wilson, was sacrificed and Thomas A. Hendricks of Indiana appointed to replace him. In December, 1855, the new commissioner issued a circular announcing the termination of the timber agencies and the institution of a new policy. Thence-

[18] *Ibid.*, 8–9, 16.

forth the duty of protecting the public timber would devolve upon the regular land office registers and receivers of the various districts. There was little danger that this new system would embarrass the administration, for the local officials were residents of the communities in which they worked, and hence were dependent upon the good will of their neighbors for their social standing and continuance in office. Moreover, they received no increase in pay for their added duties.

The instructions of 1855 ordered the registers and receivers to exact from deliberate trespassers the value of the stumpage or to seize and sell the logs illegally taken. A trespasser who had acted through ignorance, however, without intending to steal from the government, would be allowed to retain the logs if he paid the price of the land and the costs of the action. The interests of pre-emptors and other settlers were always to be protected. Fundamentally, therefore, the new policy differed from the old only in the machinery of enforcement. It proved to be highly satisfactory to the administration, for it had the great virtue of enabling the government to please the Westerners without having to abandon the principle of forest protection. In 1856 and for two years thereafter the General Land Office reported that all was quiet on the frontier and that it had received no "serious complaint" about trespass actions.

Compromises continued to be made with trespassers despite the specific prohibition of them contained in the circular of 1855. Finally, in 1860, the secretary of the interior gave his official sanction to compromises made in accordance with stated terms: payment by the trespasser for the stolen stumpage and expenses of seizure and his purchase of the land at the government price. As he explained two years later, the secretary felt that the government should be content to get the regular price of the land. In 1865 the commissioner of the General Land Office declared that the policy of compromise was proving a great success, but that rigid enforcement of the law would never be possible. He admitted ruefully, "Experience has taught us that when community interests conflict with law, and public opinion is in conflict with its enforcement, it becomes virtually inoperative." It was just such a conflict between the law of 1831 and the conditions of pioneer life, he went on to say, that had forced the land office to adopt the compromise policy. By 1877 compromise had become so common a method of settlement that local registers and receivers made a practice of it without question. Sometimes permission to cut was given even in advance of trespass. Thus again a system of permissive cutting was evolved which closely resembled the licensing plan later advocated by conservationists.[19]

After the Civil War events moved swiftly to end the free and easy execution of the trespass laws under the circular of 1855. Experience gradually convinced the land office that the proceeds from the sale of confiscated logs could not properly be used to pay the agents deputized by the registers and receivers. In 1872, therefore, Congress made what was the first direct appropriation for the suppression of timber depredation. It was a small one, only ten thousand dollars, but it was the beginning of regular congressional appropriations for the purpose. In 1874, for the first time since 1855, the report of the General Land Office failed to praise the efficiency of the compromise system. Instead it spoke of the "great waste and useless destruction" that resulted from the "unrestrained and reckless pursuit of gain."

In 1876 Joseph A. Williamson came to the land office as commissioner. Williamson had strong views on the control of timber depredations, and almost immediately recommended legislation requiring the absolute confiscation of all timber cut on the public lands, punishment of the culprits by fine and imprisonment, and the prohibition of compromise. Like some of his predecessors, however, he was convinced that no law could prevent trespass, and that the only way to save the nation's forest resources was to sell them with as much speed as possible to private individuals, who were better able than the government to protect their property. Early in 1877 Williamson ordered the local registers and receivers to render him a full statement of the

[19] Donaldson, *Public Domain*, 357; *Report of the Commissioner of the General Land Office*, 1856, p. 31; 1857, p. 24; 1858, p. 34; 1862, p. 20; 1864, p. 21; 1865, p. 26; 1866, p. 33; 1867, p. 95; 1877, p. 17–19; Ise, *United States Forest Policy*, 40; Cameron, *Governmental Forest Control*, 154–156, 161–164.

timber business done by their offices since 1855. When he received their reports he was shocked to learn that in the twenty-one-year period 1856–77 the Treasury had received in return for the millions of dollars worth of timber cut on the public domain only the net sum of $154,373.74, hardly more than the value of the timber on five hundred acres of good pineland. Convinced that the old system was useless, he asked Congress for authorization to appoint special agents who would be paid out of a definite appropriation, as had been those deputized by local registers and receivers since 1872.

In March, 1877, Carl Schurz became secretary of the interior. An immigrant from Germany, Schurz was imbued with the philosophy of forest protection that had long been traditional in his fatherland, and he proved to be sympathetic with all of Williamson's plans. The stage was set for the abandonment of the old system.

Williamson had already decided that the existing statutes empowered him to appoint agents without further authorization from Congress, and the new secretary agreed with him. He was instructed by Schurz to appoint and supervise the necessary agents, who were to report directly to the General Land Office. In May, 1877, he notified the registers and receivers that they were relieved of their timber duties. The public timberlands were divided into districts and assigned to special agents. A new policy had gone into effect.

Ignoring the temper of the West, Schurz and Williamson now instituted a new crusade against trespass. The drive was relentless and uncompromising. Before the end of the year twenty-nine civil suits had been instituted at St. Paul against trespassers in Minnesota and Wisconsin, and a federal grand jury had returned twenty-five indictments. Every effort was made to bring criminal proceedings not only against the poor men who had done the actual cutting but also against "the parties who are found to have organized and directed the stealing . . . on a large scale, and derived from that criminal practice the greatest profit." Congress encouraged the work with a steady increase in appropriations, which permitted an ever larger personnel. In

1878 the appropriation was twenty-five thousand dollars, in 1879 forty thousand, in 1882 seventy-five thousand, and in 1890 a hundred thousand. During the same period the working force was enlarged from eleven special agents to fifty-five.[20]

Wisconsin was still a fertile field for the investigation of timber depredations. By the middle of 1878 government agents had discovered trespassing involving nearly three-quarters of a million board feet of logs. During the next year the timber agents reported the theft of thirteen million feet of logs and recovered more than two million feet without legal proceedings. In many cases the land office was hampered by the fact that the stolen material had changed hands so often that it was impossible to collect sufficient evidence for conviction. Moreover, many sawmill owners who had been "parties to trespass" had agreed among themselves not to give any information to the federal agents. In several places the woods had been set afire to destroy evidence of trespass. The land office soon found that despite increasing appropriations it could not respond to all the requests that came "from every direction" to stop the plundering of public timber in Wisconsin. In 1880 three of the four agents in the state had to be withdrawn for lack of funds, although they were working on cases of "vast magnitude involving ... important interests to the government."[21]

Schurz's campaign produced in Congress a reaction almost as acrimonious as that of 1852. In May, 1879, a bill was introduced in the House to relieve trespassers from prosecution for timber depredations committed before the first of March if they would pay the regular price of the land. The proposal was favored by many members of both houses on the ground that inasmuch as previous administrations had led lumbermen to believe that trespassing would be condoned, it would be "cruel and harsh" to institute now a sudden campaign of severe prosecutions. A

[20] *Report of the Commissioner of the General Land Office,* 1874, pp. 40–44; 1876, p. 20; 1877, pp. 19–23; Hough, *Report on Forestry,* 3:8, 13; Donaldson, *Public Domain,* 358; Ise, *United States Forest Policy,* 107.

[21] *Report of the Commissioner of the General Land Office,* 1878, pp. 123–124; 1879, p. 186; 1880, p. 170; Donaldson, *Public Domain,* 358; *Northwestern Lumberman,* March 10, 1877; *Chippewa Herald,* April 13, October 12, 1877.

year later, in June of 1880, the bill passed with amendments somewhat restricting its operation.[22]

One thing the debate had revealed: Wisconsin lumbermen were no longer strongly opposed to the enforcement laws. Indeed, the state's congressmen furnished staunch opposition to what they called the "bill to license timber thieves"; and their colleagues from Michigan, Pennsylvania, and New England, all old lumbering regions, were equally active in fighting it. One need not look far to discover the reason for this change of heart. By this time most of the public timberlands had been taken up by private owners. The practice of trespassing, which had been so rewarding to lumbermen making a start in life now threatened to undermine the foundations of the institution of private property. The once-struggling pioneer lumberman had become a highly respected, often wealthy businessman who had to observe the code of ethics upon which his new status rested. He could not compete with the upstart, dishonest adventurers who still maintained the old frontier attitude. As a timber agent wrote Governor Fairchild in 1870, "All legitimate and responsible lumbermen are anxious that cutting upon the public lands should be suppressed as the practice . . . is keeping the market unsettled by thrusting upon it timber plundered from the public lands and put in market by labor obtained by fraudulent promises." Moreover, the average resident of a lumbering community had by 1880 begun to realize, at least dimly, how great an injury his section suffered from the denuding of the forest lands. Thus private interest and concern for the public welfare were both helping to strengthen conservation sentiment.

Nevertheless trespassing continued in Wisconsin. In 1881 thirty-nine cases were reported by the single federal agent maintained in the state, and the next year the General Land Office cited Michigan and Wisconsin as the "field of perhaps the most extensive trespassing in the country."[23]

Schurz's vigorous attack on the problem was not continued

[22] *Congressional Record,* 46 Congress, 2 session (1880–81), 4535–4539; Donaldson, *Public Domain,* 358; *United States Statutes at Large,* 21:237.

[23] Sam Harriman to Governor Lucius Fairchild, November 26, 1870, in Executive Records: Timber Agent; Donaldson, *Public Domain,* 1050–1051; *Report of*

by his successor, but in 1885 the campaign was renewed by Commissioner William A. J. Sparks with the encouragement of his superior, Secretary L. Q. C. Lamar. Despite repeated recommendations of the Department of the Interior that Congress pass a comprehensive act codifying the laws on timber trespass and its penalties and giving the land office clearer authority to suppress it, the century ended without any improvement. Registers and receivers lacked even the power to subpoena witnesses. In 1898 the land office admitted its inability to cope with the problem. "The best efforts of this office to wholly check such operations," wrote the commissioner, had been "unavailing."[24]

The state as well as the federal government was faced with the trespass problem. Before 1855 the public and the state government itself gave it little thought; for the protection of timber on swamplands there was no machinery at all, and for its protection on school lands only the provision that the superintendent of schools in each township was to inform the district attorney in his county if he happened to receive any information of trespass. By 1865, however, the legislature had adopted a series of regulations defining trespass and providing penalties and means of enforcing the law. In 1855 the unauthorized cutting of timber on state-owned swampland became a misdemeanor punishable by fine or imprisonment. The apprehension of trespassers was assigned to the sheriffs and constables, and district attorneys were charged with the responsibility of bringing suit against offenders to recover the value of the stolen timber. Two years later the governor was authorized to seize timber stolen from swamplands and sell it at public auction, but it was not until 1865 that trespass laws in force for other kinds of state lands were made expressly applicable to swamplands.[25]

the Commissioner of the General Land Office, 1881, p. 374; Oshkosh *Northwestern,* March 28, 1878.

[24] Donaldson, *Public Domain,* 1051–1052; *Report of the Commissioner of the General Land Office,* 1885, pp. 81–83, 307; 1886, pp. 102, 443, 445; 1887, pp. 489–494; 1898, pp. 103, 379; *Chippewa Herald,* June 16, 1882, March 12, 1886; *Eau Claire Free Press,* November 6, 1890; "Report of Chief of Division of Forestry," in *Report of the Commissioner of Agriculture,* 1885, p. 204; Ise, *United States Forest Policy,* 81, 86–88, 90–92, 106.

[25] *Report of the Commissioners of School and University Lands,* 1854, p. 27; *Revised Statutes,* 1849, ch. 134, sec. 49; *General Laws,* 1854, ch. 134; 1855, ch. 84;

Additional legislation in 1860, 1864, and 1865 provided for the appointment by the land commission of special timber agents to protect state-owned forests. Such agents were authorized to seize without legal process any materials stolen from state lands and sell them at auction; and any interference with their performance of duty or with the materials they had seized was made a misdemeanor punishable by fine and imprisonment. At the same time the penalties for trespass were increased, and the agents were forbidden to compromise with trespassers. Nevertheless the trespasser still had the privilege of purchasing the lands from which he had removed the timber if they were open to private sale. If he had taken steps to buy the land under the state's liberal credit policy before he began his depredations, he could settle his account by paying what was still due plus a penalty of twenty-five per cent of the purchase price. If he had not taken such steps, he could still buy the land, but in that event the penalty was to be fifty per cent of the purchase price. In either case the state, after collecting the full price and the penalty, dropped all court actions against the trespasser and returned all seized materials to him.[26]

Neither the early laws nor the subsequent legislation had much effect on trespassing. So widespread had the practice become by 1856, and so fruitless were the efforts to prevent it, that the governor was driven to recommending immediate sale as the only feasible way of preventing losses from trespass. The private owner, he implied, was better able than the state to protect his property. It is significant that at almost the same time that the federal government was encountering such intense opposition in its efforts to stop trespass, there should have been so few reports of violence against state officials who presumably were protecting state timber. On one occasion a group of trespassers drove off a sheriff's posse that attempted to prevent them from cutting trees on state land near La Crosse, but these trespassers were "foreigners" from Minnesota. The failure of local officials

1857, ch. 54; 1865, ch. 537; *Compilation . . . Relating to the Swamp and Overflowed Land Fund,* 19, 25, 27, 61.

[26] *General Laws,* 1860, ch. 277; 1864, ch. 233; 1865, ch. 377; *Report of the Commissioners of School and University Lands,* 1864, pp. 7, 20–22.

to stop trespassing, Governor Randall declared in 1858, was due not to impotence but rather to a "patient forbearance" that "quietly endured the spoiling of the state lands" in return for "a portion of the fruits of the evil."[27]

Twenty special timber agents appointed under the law of 1865 proved to be no more capable of coping with timber spoliation than the local officials had been. Despite the fact that in 1865 they detected trespass on more than eleven thousand acres, involving more than nineteen million board feet of lumber besides other articles, the "shameful business," reported the land commissioners, continued to be carried on by "important corporations and prominent men"—so extensively at times "as to cover the rivers with stolen logs." The agents' failure, the commissioners decided in the light of an investigation made in 1870, was the result of "corruption and mismanagement." They considered that the number of agents was excessive, though the twenty authorized by the law of 1865 had been reduced to twelve in 1870. Accordingly they recommended that the state be divided into four land districts, for each of which a single agent should be appointed. This recommendation the legislature enacted into law, at the same time doubling the penalty for trespass.[28]

For several years the commissioners reported their satisfaction with the operation of the law, but in 1876 Governor Harrison Ludington, himself a lumberman, criticized the system as one designed merely to collect stumpage value after the trespass had occurred and not to prevent lawbreaking. When he was informed by a timber agent that some trespassers paid as little as one or two dollars a thousand board feet for timber worth from five to eight dollars, the governor told the legislature that it had come to be "regarded as a profitable business to strip state lands of their timber and then effect a settlement." The legislature,

[27] *Compilation . . . Relating to the Swamp and Overflowed Land Fund,* 208; *La Crosse Independent Republican,* February 3, 1858; *Governor's Message,* 1858, p. 28.

[28] *Report of the Commissioners of School and University Lands,* 1864, p. 26; 1865, p. 343; O. R. Dahl, timber agent, to Governor William E. Smith, December 31, 1873, in Executive Records: Timber Agent; *General Laws,* 1871, ch. 21.

agreeing with the governor that the "only true policy" was one of prevention, again revised the law in 1876 by transferring the appointment of the state's timber agents from the land commissioners to the governor, increasing their number from four to six, and authorizing the governor to "guide the conduct" of his appointees.[29]

The new law, too, fell far short of accomplishing its purpose. Succeeding governors permitted themselves to be so swayed by political considerations that the machinery all but broke down completely. Prominent lumbermen with political influence took a hand in obtaining the appointment of agents favorable to themselves, and party leaders demanded appointments for their henchmen; little or no consideration was given to the real requirements of the situation. The results were appalling. Agents with no knowledge of woodcraft were stationed where no protection of timber was needed. Most of the agents, a correspondent of the governor's said in 1877, were newspaper editors who, "notwithstanding there honourable qualifications as authors and politicians . . . have ether neglected there duty or been destitute of the qualifications required to detect trespassers." Another correspondent, recommending the appointment of a friend as timber agent, wrote with engaging frankness: "He is a man of unquestioned fitness for this position. He can do more real *Solid work* for the Republican Party, than any other man . . . on the Wolf River."[30]

At the root of the situation, of course, was a public opinion that remained indifferent to or actually opposed the efforts of the state government to protect its land endowment. In 1864 the

[29] *Report of the Commissioners of School and University Lands,* 1871, p. 5; 1873, p. 3; 1874, p. 3; *Governor's Message,* 1876, p. 13; H. Borchsenius to Harrison Ludington, July 6, 1876, in Executive Records: Timber Agent; *Session Laws,* 1876, ch. 314. The new law applied only to lands "which are or have been subject to entry and purchase at the office of the commissioners of school and university lands." Since the law of 1871 was repealed, other state lands were presumably no longer protected.

[30] Letters to Governor William E. Smith from H. W. Gilkey, November 16, 1877, James H. Jones, January 17, 1877, and O. R. Dahl, November 10, 1877; Thaddeus C. Pound to Jeremiah Rusk, January 10, 1882; Philetus Sawyer to Rusk, January 7, 1882, in Executive Records: Timber Agent; *Chippewa Herald,* April 20, 1877.

state land commission appealed to the citizens of the state to "condemn by voice and conduct" the practice of stealing timber from state lands and to "pursue all means to deter and punish offenders." But a year later the discouraged commissioners were still complaining that "efforts for the detection and punishment of the criminals are thwarted and even resented as unwarrantable interference with private rights. To a considerable class, including many besides those interested, this robbery of the children's inheritance has become by long use to seem excusable and even right." In the same vein a bedeviled timber agent complained, a few years later, of the obstacles he met "in attempting to discharge my duty":

> I have to encounter not only the difficulties necessarily incident to traversing a wilderness embracing many hundred miles of territory in search of proof of the extensive depredations which have been committed upon those lands, but I am compelled to meet a powerful combination of lumbermen representing millions of capital and all having a common interest to thwart me in my efforts to bring trespassers to justice.

And he was thwarted by others than the lumbermen themselves. "The interest of the entire Community is in opposition to me and my acts in enforcing the Law," he wrote. "I hate to stand in a Community like Ishmiel with my hand against every man and every man against me." The agent was indeed "in a fearful minority."[31]

Throughout the rest of the century little improvement was made in the administration of the state trespass laws. For the most part both government and public opinion took a lackadaisical attitude toward the problem. Not until conservation had become a passion with a substantial group, toward the end of the nineties, was some improvement achieved. In 1897 the legislature set up a Forestry Commission to study the whole problem of forest conservation, and shortly after the turn of the century the office of state forester was created to consolidate all

[31] *Report of the Commissioners of School and University Lands,* 1864, p. 8; 1865, p. 343; Sam Harriman to Governor Lucius Fairchild, April 1, 7, 1869, November 26, 1870, in Executive Records: Timber Agent.

activities connected with timber. In 1906 the forester reported that as a result of the "unaccountable" leniency which had been accorded trespassers in Wisconsin for many years, hardly a single forty owned by the state or the federal government had escaped trespass. In 1905 the legislature again increased both criminal and civil penalties for trespass, and the situation began to improve. "The man who deliberately steals timber is gradually coming into his own," reported the forester in 1905, "and is looked upon as a twin brother to a horse thief." But the question was still far from settled. A new class of timber thieves was created by the land dealers who sold sandy cutover lands under the guise of fertile farms to unsuspecting settlers. After they became discouraged such poor settlers often succumbed to the temptation offered by the timber on nearby state lands. Town boards, judges, and juries were inclined to be lenient with them, inasmuch as they had no money for the payment of fines, and to put them in jail only threw the burden of supporting their families upon the towns.[32]

Private landowners were likewise the victims of trespassing. Once having formed the habit of helping himself to government timber, the trespasser found it hard to desist when the government lands had been taken up. Indeed he could not easily distinguish between privately owned and government lands. Ordinarily the depredator preferred the latter, for the private owner usually hired a watchman for his lands. Charges of trespass were usually settled out of court by an agreement between owner and trespasser, though there was of course no proscription of court action. Timber trespass on private lands was defined as larceny as well as trespass, but it was hard to prove in court that the offender knew that the trees he had cut in the unfenced wilderness were someone else's property.[33]

[32] *Report of the Commissioners of Public Lands,* 1903–04, p. 27; *Report of the Chief Inspector for the Department of Public Lands of the State of Wisconsin,* 1894–98 (Madison, 1899), 5–12; *Report of the State Forester,* 1906, pp. 18–19; 1907–08, pp. 84–85; *Session Laws,* 1905, ch. 264; *Eau Claire Free Press,* July 16, 1891; *Chippewa Herald,* July 10, 17, October 23, 1891.

[33] Warren, *Pioneer Woodsman,* 57; A. G. Pinney to Moses M. Strong, June 1, 1857, in the Strong Lumber Papers; W. H. Race to Ingram and Kennedy, April 23, 1874, and W. P. Graham to Ingram and Kennedy, June 20, 1870, in the

Timber on Indian reservations in Wisconsin constituted another temptation to dishonest persons. The government resisted most of the lumbermen's efforts to have these pinelands put on sale, but the secretary of the interior from time to time issued permits for the disposal of "dead and down" timber and the cutting of trees that could be sacrificed without harm; generally speaking, the Indians themselves were not permitted to enter the lumber business independently. Unscrupulous lumbermen, aided by conniving Indian agents, succeeded in defrauding the Indians of much of the return they should have received for their logs. Often they made a smaller profit on the logs than did the lumbermen to whom they sold them. Despite many complaints of timber stealing on the Indian lands, it was not until 1888 that Congress extended the anti-trespass legislation to Indian reservations.[34]

One must not exaggerate the moral turpitude involved in timber depredations. The temper of the frontier being what it was, trespass was almost inevitable, and both federal and state governments adopted land policies that encouraged rather than checked the practice. On the other hand, it must be said that the dishonesty that accompanied the disposal of the public domain and the failure of government officials to prevent it do not present an attractive picture. Perhaps Governor Harvey's interpretation of the situation in Wisconsin can be applied to the country as a whole: "New communities are seldom benefitted, permanently, by rich doweries of land. On this point, the ex-

Ingram Papers; Bruncken, *Forests and Forestry,* 193; *La Crosse Independent Republican,* May 2, 1855; *Session Laws,* 1881, ch. 170; 1889, ch. 397.

[34] *Chippewa Herald,* August 14, 1874, October 1, 1880, December 15, 1882, January 5, 1883, May 13, 1887, April 27, December 14, 1888, January 25, February 8, 15, December 27, 1889, May 16, 1890, February 20, 1891; *Eau Claire Free Press,* July 22, 1875, May 28, 1885, March 22, 1888, January 31, April 18, December 24, 1889, October 23, 1890; Oshkosh *Northwestern,* October 20, 27, 1870, March 2, 1871, February 13, 1879, March 2, 1882, March 15, 1888, January 31, February 14, October 10, 1889, January 7, April 1, 1893; *Chicago Times,* May 26, 27, 1894; *Letter from the Secretary of Interior* (50 Congress, 2 session, Senate Executive Document no. 128, serial 2612, Washington, 1889); *Sale of Timber on Certain Lands in Wisconsin* (52 Congress, 1 session, House Report no. 1077, serial 3045, Washington, 1892); Philetus Sawyer to Orrin H. Ingram, April 12, 1889, in the Ingram Papers; Ise, *United States Forest Policy,* 106, 112; Augustus Cole to Holt and Balcom, March 29, 1887, in the Holt Lumber Company Papers.

perience of the State is suggestive. The influence of land grants to the State, has been witnessed in the corruption of our public men—of public opinion, and of the politics of the State, far more plainly than in the advancement of any material interest by the means."[35]

[35] *Special Message of the Governor* (1862), 9.

12. The Labor Problem

CARRIED on as it was near the edge of a primeval wilderness, pine lumbering during its classic age in Wisconsin shared in the American frontier experience. At the same time it was an industrial enterprise of a type ordinarily associated with urban development. This dual character influenced profoundly the industry's personnel and the course of its labor relations.

In Wisconsin its labor was supplied largely by immigrants from the Eastern states and foreign countries. During the 1830's virtually all woodsmen were French-Canadian *voyageurs* who had abandoned the fur trade or had migrated from Canadian lumbering regions. Expert rivermen though they were, they failed to satisfy the requirements of other branches of the industry and by 1850 were being replaced by a stream of more experienced workers from New England and other Eastern states and by numbers of European immigrants. The federal census of 1880 gave the total number of workers in the woods as 3,810, of whom 1,805 were native Americans, 915 British-Americans, 473 Scandinavians, 275 Germans, 147 Irish, 89 British, and 106 natives of other countries. Most of the foreign immigrants, especially the Germans, Scandinavians, and Poles, used the money they earned in the lumber industry to buy land in Wisconsin and establish themselves as farmers.[1]

[1] Lockwood, "Early Times and Events in Wisconsin," in *Wisconsin Historical*

A large proportion of the lumberjacks, on the other hand, were transients, particularly before the Civil War. Of the sixty-five laborers who appeared on the rolls of a Menomonie firm, for instance, only twenty-three were listed the next year. The problem of recruiting labor, at least in the earlier period, was a never ending one. During the war years it was particularly acute. Some lumbermen helped their men pay for substitutes when they were drafted, but most of the shortage was filled by importing workers from Canada and by using more women and girls in the mills. After the war there was seldom a serious labor shortage, and in times of depression there was an oversupply, which prompted workmen to oppose the admission of any more foreign immigrants. But the migratory habits of the workers still presented a problem, which by 1880 had become a troublesome one. Because of the higher wages paid for railway maintenance work, in the Western wheat fields, and on the Pacific coast, it became difficult to get an experienced crew that would remain at work throughout even one season. During the prosperous years after 1900 so many lumber workers migrated to the West that some companies were compelled to draw upon the Italian and other foreign colonies in the big cities for their replacements.[2]

Some women and children were on the payrolls of the lumbering companies not only in periods of labor shortage but also in normal times. Ordinarily their labor was used in shingle mills and other subsidiary activities rather than around the principal sawmilling machinery, though there are records of a few women working even as cruisers and lumberjacks. The proportion of female and child labor was small. In 1885 only about one and a quarter per cent of the total were women and about one-half

Collections, 2:141; Usher, *Wisconsin,* 1:179; Vinette, Reminiscences (MS.), 1; Stephenson, *Recollections,* 87, 105–106, 109; Kate A. E. Levi, *How Wisconsin Came by Its Large German Element* (Madison, 1892), 36; Pay Orders, 1865, in the Ingram Papers; Albert H. Sanford, "Polish People of Portage County," in *Proceedings of the State Historical Society of Wisconsin,* 1907, pp. 264–265; Charles M. Scanlan, "History of the Irish in Wisconsin," in the *Journal of the American-Irish Historical Society,* 13(1914):238–239; *Report of the Bureau of Labor and Industrial Statistics,* 1883–84, p. 38; Pay Rolls, 1903–1911, in the Holt Lumber Company Papers.

[2] Abraham Brawley to Moses M. Strong, May 11, 1848, in the Strong Lumber Papers; Schafer, *Winnebago-Horicon Basin,* 273; Fish, "Phases of Economic His-

per cent were children less than fourteen years old. Until 1898, when their laboring day was restricted by law to eight hours, women and children worked just as long as the men. In 1899 the legislature prohibited altogether the employment of children under fourteen in factories of any sort. As a result of such regulatory laws, women and children in the lumber mills had declined to a negligible number by 1908—a fraction of one per cent.[3]

Until about 1900 the lumber industry in Wisconsin employed, in the woods and in the sawmills, more laborers than any other manufacturing industry in the state. From a little over two thousand workers in 1860 the number increased steadily to a peak of more than twenty thousand thirty years later. In 1889 more than twenty-five per cent of all industrial workers in Wisconsin were employed by lumber companies, and in 1893 slightly more than twenty-one per cent. Railway shops, the next largest employer of factory labor in 1893, had on their payrolls only about six per cent of the total in all industries.[4]

As the largest employer of labor, the lumber industry at its height paid more in wages than any other industry. From 1888 to 1893 this expenditure approximated seven million dollars annually except in 1890, when it decreased to four million. In 1889 lumbering wages represented nearly a fourth of all wages paid in Wisconsin. But since most of the employees were unskilled laborers and their work was seasonal, the per capita wage was much lower than that of the other industries. From 1888 to 1891, for example, the annual per capita wage in the lumber industry ranked forty-sixth among Wisconsin's sixty-

tory of Wisconsin," in *Proceedings of the State Historical Society of Wisconsin*, 1907, p. 206; letters to Orrin H. Ingram from B. J. Cooper, February 4, 1865, W. W. Lea, March 20, 1865, and Rice Lake Lumber Company, February 9, 1900, in the Ingram Papers; *Dunn County News*, January 10, 1918; *Eau Claire Free Press*, March 3, 1881, October 18, 1883, February 17, 1886, September 29, 1887, May 24, 1900; *Eau Claire Leader*, March 9, 1916; *Chippewa Herald*, January 12, 1899; *Report of the Bureau of Labor and Industrial Statistics*, 1885–86, pp. 420, 429, 437; 1887–88, p. 13; 1905–06, pp. 1361, 1376, 1379.

[3] *Eau Claire Free Press*, October 20, 1887; *Report of the Bureau of Labor and Industrial Statistics*, 1885–86, p. 485; 1888–89, p. 139; 1899–1900, p. 30; 1909–10, p. 997.

[4] State Board of Immigration, *Statistics Exhibiting the History, Climate, and Productions of the State of Wisconsin* (Madison, 1867), 17; *Report of the Bureau*

WHi(X3)18377

Crandon(?), Wisconsin, probably 1910–1915. Blacksmith shop in a lumber camp. Notice the saw-holder being used by the man preparing to file a cross-cut saw blade. Photo probably by Anderson.

WHi(X3)30785

Clam River, Wisconsin, 1901. Lunch carrier looking for a driving crew along the river. Photo probably by Charland.

three industries. In 1897, when the average per capita wage for the state as a whole was $416.79, the average in the lumber industry was only $386.09. Although wage rates naturally varied somewhat from year to year, the range was not wide in the quarter century 1880–1905. Common laborers in the woods received from fifteen to twenty-five dollars a month and board, and skilled workers from thirty to fifty dollars, depending upon general business conditions and the state of the labor market. Log drivers received from two to three dollars a day and meals. In the mills head sawyers were paid a daily wage of three to four dollars, and ordinary laborers about a dollar.[5]

Seldom, however, did the workman receive all his wages in cash. Because lumbermen usually had most of their cash tied up in operations each season, they were forced to give their workmen due bills, which were usually payable in the fall or spring months, when lumber or logs had been converted into cash. Due bills were receivable for goods at the company store, and employees took part of their pay in kind. Lumber companies encouraged this practice by extending liberal credit, and since the company store was the only place where employees could buy on credit during the long payless months, most of them had no choice but to trade there. Not all companies used the due-bill system, but even those who did not do so required their employees to wait until the close of the season for their pay.[6]

The due-bill system was remunerative for many companies. Few laborers could exist for three months without some ready cash, even with liberal credit at the company store. To ease them over this difficult period the employer was almost always willing

of Labor and Industrial Statistics, 1888–89, p. 126a; 1891–92, p. 138a; 1893–94, p. 190a; 1899–1900, p. 418. These statistics do not include workers in sash, door and blind factories or planing mills, which the Bureau groups as a separate industry obtaining its materials from the sawmills.

[5] *Report of the Bureau of Labor and Industrial Statistics,* 1886–87, pp. 93, 95, 226–227; 1888–1889, pp. 139, 154; 1891–92, pp. 83–85, 106–108; 1893–94, p. 107; 1897–98, pp. 589, 591, 828; Holt and Balcom Time Book, 1866–67, and Holt Lumber Co. Pay Rolls, 1903–1911, in the Holt Lumber Company Papers; Statement of Account, dated April, 1850, and S. J. Carpenter to Moses M. Strong, January 24, 1859, in the Strong Lumber Papers.

[6] Charles Wilson to Moses M. Strong, August 31, 1852, in the Strong Lumber Papers; B. E. Reid to Orrin H. Ingram, April 1, 1875, in the Ingram Papers; *Eau Claire Free Press,* September 18, 1879.

to discount their due bills for ten per cent or more. For many years the bills could be sold only to the employer, but in 1890 the legislature made them negotiable and gave the buyer a lien on the logs which the holder of the bills had helped to cut. This did not, however, save the workman from the loss of ten or twenty per cent of his wages when he sold his time check. Although lumbermen were specifically exempted from the statute providing that wages must be paid weekly or bi-weekly in cash, they were required to furnish written evidence of their indebtedness to workers in the form of negotiable paper payable at the main office of the company or at any bank in the state.[7]

Some lumber companies adopted special devices to force employees to trade at their stores. Where the due-bill system was in use, further coercion was seldom necessary, at least until the employee had sold his bill, but elsewhere other means were developed. The simplest was a written agreement, which every employee was required to sign when he was hired, that he would buy everything he needed at the company store. In the eighties N. C. Foster and Son of Fairchild issued a passbook to each employee on the inside cover of which was printed: "*Notice*—It is expected that all persons employed by me will purchase their Goods at my Store, which is kept for the purpose of supplying employees of this Institution. I must have the Trade of those I furnish employment." The employee was required to sign the statement, "I understand the above conditions and hereby agree to the same." The passbook contained a ticket that could be punched for the correct amount each time a purchase was made.

In isolated districts where a company store was a necessity, its monopoly was secure, but many workmen were penalized by trading in so restricted a market. The price of an item was sometimes ten or fifteen per cent higher than in comparable stores in the same community. In 1885 the millworkers at Marinette organized a union and refused to trade at the company stores, thereby forcing some of the lumber companies to discontinue

[7] *Report of the Bureau of Labor and Industrial Statistics*, 1899–1900, pp. 34, 41; *Statutes*, 1898, sec, 1729a; *Session Laws*, 1899, ch. 221; *Chippewa Herald*, April 4, 11, 1890; *Eau Claire Free Press*, October 30, 1890; *Ashland Commonwealth*, January 12, 1895.

their retail stores and others to "sell at city prices" in order to keep their customers. In 1875 several lumbermen were indicted by the federal grand jury at La Crosse for issuing scrip payable in goods at their stores, found guilty of violating the federal banking laws, and fined one hundred dollars apiece. Two years later, however, the Supreme Court, acting on a similar case appealed from Michigan, decided that such scrip was legal. But the system was unpopular and by 1880 many lumber companies were abandoning it.[8]

Some companies included in their labor contracts extremely severe provisions for maintaining discipline among their employees. The contract drawn by Knapp, Stout and Company in the eighties contained, in addition to a statement of the monthly wage of the employee and the exact period for which he was hired, the stipulation that he would not be paid until the "end of the time hired for," and if he quit earlier he would receive twenty per cent less than the sum agreed upon, and not that until the end of the term. Breach of contract, which was interpreted as any violation of the rules of the company, automatically ended the employment of a worker, who again must wait until the end of his term before receiving any money. The company rules were expressed succinctly:

> Any man refusing to perform any labor he is required to do, if capable, including driving logs, will be considered to have broken his contract, unless he was hired to work at some other employment only.
>
> Drunkenness, or the introduction of intoxicating liquors on the premises, will be treated as a breach of contract in all cases.
>
> Disorderly or riotous conduct, or the wilfull or malicious damage or destruction of property, will be held a sufficient cause to declare the contract broken.

Such a contract served the company for a variety of purposes, not the least of which was the control of labor disputes. "Dis-

[8] *Report of the Bureau of Labor and Industrial Statistics,* 1883–84, p. 118; 1887–88, pp. 91–92, 111–112; *Chippewa Herald,* October 8, December 24, 1875, July 19, 1878, March 26, 1880; *Eau Claire Free Press,* November 1, 1888; La Crosse *Republican and Leader,* December 25, 1875; U. S. *v.* Van Auken, 96 U. S. 366 (1877).

orderly or riotous conduct," in the language of the time and place, meant "strike." Men who took part in a strike, therefore, automatically forfeited twenty per cent of their wages and had to wait perhaps several months before getting the balance. Meanwhile their credit at the company store was cut off. The company abided rigidly by the terms of its contracts. Men who were hired in the spring were not paid until the closing of the mills in November; those who were hired in the fall to work in the woods received no pay until the logs came down in the spring. The company usually advanced enough money for the payment of house rent and doctor bills, but not enough to buy any food or clothing. Employees therefore had no choice but to trade at the company store, where even clothing was about five per cent higher than at other stores. In 1886 an employee who wanted to buy a cow had to borrow the money from the company and pay interest, though he had more than enough due him in wages. When a Norwegian lumberjack was killed in the woods while on his job, his wages were docked twenty per cent because he had failed to complete his term![9]

A series of lien laws passed by the state legislature after 1860 protected the right of the laborer to collect his wages. Thus anyone working for the lumber industry was entitled to a first lien on the material he had helped to produce. By recording it in the office of the proper district lumber inspector he could validate it fully before the law. In Chippewa County it was sufficient merely to record such a claim in the office of the clerk of the district court. Every stockholder in a lumber company was personally liable for the wages of its employees up to the value of his stock or the amount of six months wages.[10]

Although the average worker in the lumber industry had many grievances, unionism was introduced only slowly. In Eastern industrial centers trade and labor unions had been developing since the beginning of the nineteenth century, but in the sawmilling centers of the Midwest labor did not engage in group

[9] *Report of the Bureau of Labor and Industrial Statistics*, 1887–88, pp. 111–113.

[10] *Ibid.*, 1899–1900, pp. 45, 69, 75; Hagenah, *Law of Labor*, 121–126; *Statutes*, 1898, ch. 84, sec. 1739; ch. 85, sec. 1769; ch. 143, sec. 3329, 3342a.

action on any extensive scale until a decade and a half after the Civil War. For this delay there were several reasons. Trade-union psychology is basically a feeling of group solidarity in the face of limited economic opportunity. Fundamentally, the motive that prompts a laborer to join a union voluntarily is the desire for security and a higher standard of living through equitable distribution of existing opportunities. The ranks of the lumber workers, however, were dominated by the frontier psychology of optimism and a sense of unlimited opportunity. Because the lumber industry was, so to speak, an island of industrialism in the broad sea of the frontier, the lumberman could hire workers who felt none of the economic restrictions that weighed on the laborer in the older industrial centers. The lumber workers, many of whom were pioneer farmers who spent several months each year in the wilderness, had the same roseate visions of the future as other frontiersmen. Most of them had come West to take advantage of the boundless opportunity it offered. They had all seen men rise from humble beginnings to positions of wealth and influence. What if the worker did suffer from restrictions today? Would he not be independent tomorrow?

Even if he did not experience precisely this glow of optimism, the worker would have found it difficult to maintain other ideas in the face of perpetual indoctrination by the more articulate members of the community. Newspapers, many of them owned or controlled by lumbermen, spoke again and again of the wealth of opportunity open to those who were willing to work. Any suggestion of complaint from labor was met with the charge of ingratitude. The sawmills had given the worker a living and a home. How could he speak unkindly of them? From the pulpit he was advised to be content with his lot. At the height of the resurgent labor movement of the eighties an Eau Claire minister preached on "The Justice of the Complaints against the Present Industrial System," using as his text a verse from *Proverbs:* "He that handleth a matter wisely shall find good, and whoso trusteth in the Lord, happy is he." Another clergyman declared that the laborer must be disabused of the idea that corporations grow richer and the poor poorer. Labor should not hate the rich, for it too had the opportunity to rise "to that condition."[11]

The promise of future prosperity bore fruit for a number of laborers. Many became homeowners on an annual wage of less than three hundred dollars. Lumber companies often encouraged their employees to build homes by selling them lumber at reduced prices or by supplying the lumber and allowing them to "work it out." Thus a number of factors combined to make the laborer in the Wisconsin lumber industry an individualist rather than a unionist in spirit.[12]

The lumberjacks were never successfully organized in Wisconsin before 1900, but sawmill workers, many of whom were skilled artisans and whose jobs were more like those in other industries, began to manifest union sentiments after 1870. The first upheaval was born of the question of working hours. Before 1880 the ordinary workday in a sawmill was from eleven to thirteen hours long, and some companies developed an elementary speed-up system by encouraging the crews of various mills to compete for the highest production of the day. When sixty workers at La Crosse went on strike for a ten-hour day in 1873, they were immediately black-listed and replaced by other men, but the reductions in wages brought about by the ensuing depression caused more strikes, many of which were accompanied by much violence. One of the most effective devices of the lumber saboteur was to drive a heavy spike into a saw log, which could not then be run into the saw without endangering the lives of the sawyers. In 1880 at the request of many millowners a bill was introduced in the Assembly which, when finally passed, increased the penalty for spiking a log to a five-year term in the state prison or a five-hundred-dollar fine. A few years after the La Crosse strike the Chippewa Valley Workingmen's Association—one of the first lumber workers' unions—was organized to work for shorter hours and the payment of wages in cash.[13]

A strike at Eau Claire in 1881 proved to be the most serious

[11] *Eau Claire Free Press,* May 28, 1885, July 8, 1886.

[12] *Report of the Bureau of Labor and Industrial Statistics,* 1885–86, pp. 443–445; 1887–88, p. 25.

[13] La Crosse *Republican and Leader,* August 2, 1873, June 17, 1876; *Chippewa*

labor dispute the lumber industry was called upon to face in Wisconsin. At the opening of the sawing season the millowners had hired their crews as usual, placing most of them under specific contract to work a twelve-hour day. In July, however—"incited," as state officials later reported, "by a few small politicians and some aggressive leaders of their own"—the employees of all the local mills demanded that their day be reduced to ten hours. When the owners refused, the men walked out. The employers took the position that, although the men were under contract to work twelve hours a day, they were willing to pay off any workman who would rather quit than work that long. Other men were hired to replace the strikers.

Reacting violently to the use of the "scabs," the striking laborers drove them from the mills, extinguished the fires in the boiler rooms, stopped all the machinery, and seriously damaged some of it. The city was "possessed" by the strikers, most of whom were said to be single men "without property interests in Eau Claire, or permanent homes anywhere." They defied the city authorities and rode roughshod over a sheriff's posse. Alarmed for the safety of life and property, the mayor asked Governor William Smith to send the state militia to Eau Claire. After a personal examination of the situation, the governor responded by despatching some three hundred soldiers to the town. There they were met by two thousand strikers, "large, horny-handed and resolute fellows, accustomed to hard work and hard usage, and fearless of danger," armed with guns, clubs, crowbars, and mill tools. "Demagogues" incited the strikers to attack, but somehow a clash was averted. The militia took control of the situation, and the mills soon resumed operations on a twelve-hour

Herald, May 26, 1876; Oshkosh *Northwestern,* August 2, 1877; *Northwestern Lumberman,* July 26, 1879; *Eau Claire Free Press,* October 16, 1879, February 5, 1880, June 30, 1881. There were some indications of labor solidarity before 1870. During the Civil War the lumber workers included "a large element" of Copperhead voters who feared that emancipation of the slaves would demoralize the labor market of the North. In 1865 Mississippi River pilots were banded together in an organization probably designed to keep wages up. Stephenson, *Recollections,* 156; Ethan Allen to Orrin H. Ingram, March 17, 1865, in the Ingram Papers. An interesting advertisement, apparently paid for by a group of lumberjacks, appeared in the *Chippewa Falls Democrat* of April 27, 1871. It condemned a certain lumberman for charging the men in his camps fifteen cents apiece for soap, "fixing the price according to the lousiness of the men."

basis. Shortly thereafter, however, most of the mills in Eau Claire went on an eleven-hour day.[14]

The strike revealed that there was considerable pro-labor sentiment throughout the state. An Oshkosh paper commented editorially, "If a laborer has no legal right to injure the person or property of his employer, on the other hand, an employer has no moral right to propose or inflict an injustice on the laborer, merely because he has the power to do so." Democratic journals almost unanimously condemned the Republican governor for calling out the militia. Any workman had the right to quit, they said, when he was no longer satisfied with the terms upon which he worked. The millowners replied that, while a workman might quit, he had no right to force others to leave their jobs. The Bureau of Labor and Industrial Statistics reported that the strikers had the sympathy of the public until they started to riot. "The lesson taught by this strike to the lawless classes, who are never union men or workingmen, was salutary. . . . The state possesses abundant power to protect both [property and life], which responds quickly to the command of the governor."[15]

After the Eau Claire "sawdust war" labor stoppages among lumber workmen became common. In the camps the men sometimes struck for butter or some other article of food they wanted. In 1885 and 1886 the Bureau of Labor and Industrial Statistics reported nine strikes—four for higher wages, four for the ten-hour day, and one for better conditions in a company boardinghouse—which were all unsuccessful. Nevertheless, in 1886 Eau Claire workmen forced a sawmill to remain on the ten-hour day when it contemplated lengthening hours once more. On the Chippewa River driving contractors were forced to increase the daily pay of workers from a dollar to a dollar and a quarter. In Marinette millworkers won the ten-hour day and cash wages, and a little later some ninety millowners in other parts of the state adopted the ten-hour day, though about a hundred and

[14] *Report of the Bureau of Labor and Industrial Statistics*, 1883–84, pp. 151–153; *Eau Claire Free Press*, July 21, 28, August 4, 1881; *Chippewa Herald*, July 22, 1881.

[15] Oshkosh *Northwestern*, July 28, 1881; *Chippewa Herald*, August 12, 19, 1881; *Report of the Bureau of Labor and Industrial Statistics*, 1883–84, p. 153.

twenty mills remained on the eleven-hour schedule. In 1884 the legislature of Michigan passed a law providing for the ten-hour day in all mills and factories unless a written contract stipulated otherwise. When in the following year the millowners of northern Michigan attempted to impose upon their workmen such contracts for the longer day, they precipitated a strike in which their employees were aided by Wisconsin workmen across the border. Neither side could be forced to capitulate, and the strike ended in a general lockout. Many mills remained closed for the season, and at those which did resume operations the employees accepted defeat.[16]

Despite frequent strikes, there seems to have been no strong organization among lumber workmen before 1885.[17] But union sentiment was growing. Early in 1885 an effort was made to form an organization of unemployed workmen at Eau Claire for the purpose of running a cooperative boardinghouse. In the same year many chapters of the Knights of Labor were organized among lumber workers, to the great alarm of employers. "The crisis [is] near at hand," one lumberman wrote his partner, "when this great question of Capital and Labour will have to be settled, and if Labour should triumph then we will have to go out of business and divide what we have with them."[18]

As the labor movement gathered momentum in the lumber industry during the early eighties, the question of labor and capital in all its aspects became the topic of the day. Questionnaires sent out by the state Bureau of Labor and Industrial Statistics showed that almost all lumber workers were in favor of forming unions, sending laboring men to the legislature and

[16] *Chippewa Herald*, July 25, 1884; *Eau Claire Free Press*, May 14, June 4, 1885; La Crosse *Chronicle*, December 17, 1885; *Report of the Bureau of Labor and Industrial Statistics*, 1883–84, pp. 153–154; 1885–86, pp. 238–247, 310–311. The Marinette mills and ninety others throughout the state adopted the ten-hour day in 1886. *Ibid.*, 1885–86, p. 245.

[17] The list of labor organizations given in the 1883–84 report of the Bureau of Labor and Industrial Statistics includes no unions of the lumber industry, not even chapters of the Knights of Labor.

[18] *Eau Claire Free Press*, January 1, 1885, January 28, 1886; *Chippewa Herald*, January 9, 1885; Oshkosh *Weekly Northwestern*, December 3, 1885; D. M. Dulaney, Sr., to Orrin H. Ingram, May 29, 1885, and Jeremiah M. Rusk to C. A. Chamberlin, May 12, 1886, in the Ingram Papers.

to Congress, and restricting immigration. The eight-hour day, too, was beginning to be agitated. Millowners charged that it would ruin the country. "You might as well try to establish by law the price of corn or wheat," said one. "The laborer working eight hours per day," wrote another, "leaves him sixteen for rest. Too much rest leads to idleness, which means going down hill." Most millowners opposed all profit-sharing plans, and almost two-thirds of those questioned were likewise opposed to the arbitration of labor disputes.[19]

By 1886 unionism had reached its height among lumber workers. The Knights of Labor had become strong enough to "investigate" millowners who discharged employees for alleged union activities. After that year, however, two factors combined to check the power of the Knights: the Haymarket incident in Chicago, which gave the enemies of unionism an opportunity to charge the entire labor movement with anarchistic tendencies and advocacy of violence, and the weaknesses of the policies pursued by the Knights of Labor.[20]

But the labor upheaval of the 1880's had important results. In the first place, it gave the lumber workers a definite program. Almost universally they demanded the ten-hour day and cash payment of wages, and in many places they succeeded in imposing their demands upon the millowners. Gradually they added other proposals: the abolition of convict labor, the regulation of child labor, a high tariff on lumber, better safety devices in sawmills, employers' liability for injuries sustained by workmen, revocation of the railway land grants, governmental control of railway and telegraph lines, popular election of senators, and repeal of the national bank act. Many of these proposals, which would have had no immediate effect on the lumber workers, were probably adopted without change from the legislative program of the Knights of Labor, but on the points

[19] *Report of the Bureau of Labor and Industrial Statistics,* 1885–86, pp. 229, 236, 355–356, 414, 432, 438–439.

[20] *Eau Claire Free Press,* March 4, 11, April 8, August 19, 1886; *Chippewa Herald,* July 16, 1886; Hotchkiss, *Lumber and Forest Industry,* 309; W. H. Day to Orrin H. Ingram, November 5, 1887, in the Ingram Papers; *Report of the Bureau of Labor and Industrial Statistics,* 1887–88, p. 40.

directly influencing his own welfare the Wisconsin worker had strong convictions.[21]

Although the Knights of Labor, the first movement for the organization of lumber workers in Wisconsin, began to decline after 1886, it was not dead, nor was unionism. Employees' organizations continued to plague the millowners with strikes as long as the industry lasted. As late as 1892 the Knights led a strike in La Crosse against the discharge of employees for union activities and obtained the promise of all millowners in the city to discontinue the practice. The Knights were also active in local politics. In the meantime the American Federation of Labor was growing in strength. During the 1890's a widespread movement developed for the formation of local rivermen's and millworkers' unions affiliated directly with the Federation. In 1890 an attempt was made to set up a national millworkers' union, apparently without success. In the mid-nineties, when the Populist Party was at its height, many local millworkers' unions supported it. Before 1900 the ten-hour day had become almost universal, and by 1905 the deferred payment of wages was no longer common.[22]

On their side, the millowners presented a more or less solid front against the organized strength of labor. The iron hand of competition was encased in a velvet glove when concerted action was needed in a labor dispute. At the first sign of trouble the lumbermen's exchange in a sawmilling center was converted into an employers' organization. During the La Crosse strike of 1892, for instance, the local Lumbermen's Exchange declared a lockout in all the city's mills. By threatening to shut down their mills permanently the lumbermen could always get the active support of small businessmen, who would try to induce the la-

[21] *Ibid.*, 1887–88, pp. 60, 70, 93.

[22] La Crosse *Weekly Chronicle*, April 28, May 12, 1892; *Chippewa Herald*, April 29, May 13, 1892, July 27, 1894; *Eau Claire Free Press*, May 14, 1889, July 17, 1890, August 18, 25, September 29, 1892, August 2, 1894; Chippewa Falls *People's Paper*, August 18, 1894; *Ashland Commonwealth*, November 3, 1894; Milo S. Peyton to C. A. Chamberlin, July 26, 1890, and W. H. Bundy to Orrin H. Ingram, August 10, 1899, in the Ingram Papers; Augustus Cole to Holt and Balcom, April 6, 1887, in the Holt Lumber Company Papers; *Report of the Bureau of Labor and Industrial Statistics*, 1887–88, p. 93; 1891–92, pp. 122–123; 1905–06, p. 1327.

borers to return to work. Labor contracts stipulating that a man forfeited part or all of his wages if he quit his job before the end of a specified time were used as clubs to bring dissatisfied workmen to terms. Large companies with many mills could step up production schedules in those which were unaffected when a strike closed down one of them. Despite their natural antipathy to unionism, however, some lumbermen came to accept willingly their workers' demands for cash wages and the ten-hour day, having discovered that contented labor could produce as much lumber in ten hours as dissatisfied workers could produce in eleven.[23]

By 1900 the state legislature had adopted a series of laws that constituted an extensive labor code. Both boycotting and blacklisting were declared illegal, and discrimination either in favor of or against union members was forbidden. Unlawful assembly, riot, and conspiracy were made criminal offenses. In 1899 the legislature created a board of arbitration with power to deliver binding decisions in certain types of disputes. A few months after its creation the board succeeded in settling a millworkers' strike at Rice Lake.[24]

A large part of the labor code was devoted to problems arising from the physical hazards of the industry. There can be little doubt that logging and sawing were among the most dangerous of occupations. Accidents took a yearly toll of life and limb in the woods and on the drive, as did the sawmill machinery unequipped with safeguards. Nor was it the "green hands" who suffered most of the injuries in the mills, for they respected the buzzing saws and whirring machine belts; rather it was the old, experienced men who had grown careless of danger. Injuries ranged from the loss of limb in the saws to loss of life from the explosion of a steam boiler. Flying pieces of wood thrown

[23] Letters to Orrin H. Ingram from J. W. Hendrick, July 9, 1894, G. M. Pullman, February 16, 1895, Rice Lake Lumber Co., August 11, 1899, and Charles Horton, August 12, 21, 1899; Rice Lake Lumber Co. to Empire Lumber Co., August 15, 1899; Empire Lumber Co. to D. Bronson, August 12, 1902, all in the Ingram Papers; La Crosse *Weekly Chronicle,* April 28, 1892.

[24] *Report of the Bureau of Labor and Industrial Statistics,* 1899–1900, pp. 41, 99–101; *Statutes,* 1898, ch. 182, sec. 4466a, 4466b; ch. 184; ch. 332; *Eau Claire Free Press,* August 24, 1899.

out by a saw often caused painful wounds. It was said that an experienced shingle sawyer could always be recognized "by the loss of his fingers and the way they are cut." Indescribably horrible injuries were inflicted upon the occasional worker who fell upon a moving circular saw. Serious accidents occurred with alarming frequency. In 1885 the Menominee River Hospital at Marinette treated thirty-five mill-accident cases for an average of more than thirty days each, and in the same year an Ashland hospital treated thirty-three millworkers for an average of twenty-one days. Of the five hundred and sixty-five accidents reported by the Bureau of Labor and Industrial Statistics in 1907, almost twenty per cent had occurred in the woods.[25]

Until the seventies most sawmill towns had no adequate facilities for treating these casualties, but where hospitals were available several forms of hospital insurance had been developed by 1880. In 1881, when woodsmen collected funds for the building of a hospital at Chippewa Falls, each contributor received a ticket entitling him to "all the privileges the house affords." Eau Claire had a Hospital Aid Association in 1886. After 1890 private insurance companies sold hospital tickets to lumber workers. A ticket costing about twelve dollars entitled a workman to all the medical and hospital care he needed during the course of a year.[26]

Under the common law and the statutes of Wisconsin an employer was liable for damages if an employee was injured as a result of his negligence, provided the claimant could prove that his injuries were not the result of his own carelessness or that of

[25] *Eau Claire Free Press,* May 2, 1867, March 26, 1868, February 11, 1869, November 18, 1880, January 11, 25, July 19, 1883, May 6, 1892; Oshkosh *Northwestern,* June 29, 1860, November 9, 1865, August 11, 1894; *Chippewa Falls Democrat,* May 26, 1870, February 23, April 13, 1871; *Chippewa Herald,* February 11, 25, July 29, 1871, January 1, 1875, January 26, 1877; *La Crosse Leader,* February 11, 1871; La Crosse *Republican and Leader,* May 16, 1874, September 16, 1876, May 24, 1879; La Crosse *Weekly Chronicle,* July 21, 1881, February 21, 1889; *Report of the Bureau of Labor and Industrial Statistics,* 1885–86, p. 246; 1887–88, p. 106; 1907–08, pp. 23, 39.

[26] *Chippewa Herald,* December 2, 1881, September 22, 1882, January 26, 1883; *Eau Claire Free Press,* March 14, 1886, February 18, 1897; *Milwaukee Journal,* December 4, 1938; C. H. Bunker to Orrin H. Ingram, November 26, 1895, in the Ingram Papers.

a "fellow servant." An employer was bound to furnish a reasonably safe place to work, and he could not plead assumption of risk by the employee unless the machinery had been properly equipped with safeguards. Since sawmills were rarely so equipped before 1890, heavy damages were sometimes collected by injured employees, such as the millworker who in 1889 collected five thousand dollars from the Menominee River Lumber Company as compensation for the loss of a leg. Many of the millowners became "somewhat weary," as one of them confessed in 1890, "of having our employees look to us in case of accident for compensation," and an increasing number took out liability insurance. It cannot be said, however, that most injured men received adequate compensation according to present-day standards. Usually the employer could plead assumption of risk or negligence on the part of the employee or negligence of a fellow worker. Most injured men were willing, even if they had a good case, to accept a small amount from their employers rather than go to court. When the state set up rigid safety requirements for factories during the nineties most millowners were forced to install safety equipment. In 1891 and 1892 state factory inspectors ordered the installation of one hundred and fifty-six safety guards on circular saws, and each year thereafter the number of safety devices in sawmills increased. The result, of course, was a decline in the number of injuries to workmen. A large Oconto firm employing more than a hundred men in its mill listed over the two-year period 1908–10 only eighty-eight accidents, virtually all of which had inflicted only minor cuts or bruises.[27]

[27] *Report of the Bureau of Labor and Industrial Statistics,* 1891–92, p. 145a; 1893–94, p. 199; 1907–08, p. 76; 1911–12, p. 105; *Chippewa Herald,* May 17, 1889; *Eau Claire Free Press,* March 16, 1893; W. H. Day to Orrin H. Ingram, October 10, 1890, in the Ingram Papers; Record of Accidents, March 31, 1908, to April 10, 1910, in the Holt Lumber Company Papers.

13. Political and Social Influences

WITHIN a generation after its birth in Wisconsin the rugged young lumber industry had attained a dominant position not only economically but politically and socially as well. After 1870 the business leaders of the industry—the Thorps, Sawyers, and Stephensons—exercised the political influence to which the magnitude of their enterprises seemed to entitle them; many of them managed, too, to establish themselves among the state's "society" leaders. But in neither their public nor their private lives were they distinguishable from the successful tycoon elsewhere in the nation. They conformed to the general pattern of middle-class life in America, and whatever permanent influence they had was blended into the social fabric as a whole. The shanty boy and lumberjack, on the other hand, introduced a distinctive way of life that has left a deep imprint on the state's lumbering sections.

As elsewhere in the nation, the man of wealth and position in Wisconsin could exert an influence on government in several ways, most of which required the judicious expenditure of money in the right places. "G—d d—n it, money counts in Chippewa County!" a candidate for Congress was reported to have exclaimed in 1875 by way of expounding how to get votes.[1] He

[1] *Chippewa Herald,* October 29, 1875.

might have added that the votes did not need to be bought outright, although they often were; quite as effective in assuring the results of elections was the use of money to oil the local political machines or to procure the support of newspapers. Nor was its use limited to election campaigns. It was of inestimable value in lobbying for the industry and, on occasion, in bribing legislators and administrative officials.

The great lumbermen of Wisconsin, for whom rich prizes were at stake, made full use of the power they enjoyed by virtue of their position. They made generous contributions to one or the other of the two political parties, and many of them ran for office. In 1880 Philetus Sawyer's enemies accused him of having bought outright much of his support for the United States senatorship, and a quarter of a century later Isaac Stephenson, according to his own admission, spent more than a hundred thousand dollars to win the nomination for the same office. Some lumbermen owned newspapers, and others obtained at least partial control of political journals by carrying large advertising accounts. In the later years of the century certain millowners were accused of having imported at their own expense unnaturalized aliens whom they could send to the polls to vote for the "right candidates" in important local elections. Ordinarily laborers did not need to be bribed to vote as the company desired. Some employers asked the prospective employee what his political faith was or threatened to close their mills if opposition candidates were elected. The nomination of friendly candidates for all offices was facilitated by the caucus system, which could be fairly easily controlled by men of influence.[2]

Local government, as Governor Harvey pointed out in 1862, was particularly susceptible to the influence of the lumbermen.

[2] *Governor's Message,* 1875, p. 11; Cephus Shekell to Moses M. Strong, October 4, 1850, in the Strong Lumber Papers; W. D. Allen to Orrin H. Ingram, July 20, 1875, and P. H. Swift to Ingram, January 22, 1889, in the Ingram Papers; *Chippewa Herald,* September 22, November 24, 1876, August 28, 1896; *Eau Claire Free Press,* December 30, 1880, November 1, 1888, March 23, September 28, 1893; Nelligan, *Life of a Lumberman,* 180; *Report of the Senate Members of the Joint Senatorial Primary Investigating Committee* (Madison, 1911), 6. The best biography of a man who combined politics with lumbering is Richard N. Current's *Pine Logs and Politics: A Life of Philetus Sawyer.*

Before any governmental machinery had yet been established in the wilderness they were carrying on logging operations there and performing, through the camp foreman or superintendent, many of the functions of local government. After counties and towns were organized they continued to carry on the duties of government, now through official agencies. Isaac Stephenson, one of those who as officeholder was charged with many of the same duties he had once performed as superintendent of logging camps and sawmills, wrote that "from the building of roads, the carrying out of public improvements, and the regulation of general affairs as manager of the mill, it was but a short step to membership on the county board." His duties he accepted as "a matter of course, as I would have done them whether I had held office or not." The lumbermen's control of county and town governments was manifest in the number of political positions they held and the number of concessions they won from their communities.[3]

Newspapers commonly referred to certain counties as belonging to this or that large lumber company. In the mid-seventies, for example, the Dunn County delegates to a state senatorial nominating convention were dubbed "Knapp, Stout and Company's hogs," and a local politician, openly boasting of his influence with that company, boldly proclaimed in Chippewa Falls that Barron County voted as he wanted it to vote. Thus the lumbermen found it not too difficult to seat sympathetic lawmakers in the legislature. Not infrequently a lumberman held a state office himself, but a friend in office could serve him just as well. "If I can do you any good here, any way, let me know," an assemblyman wrote a lumberman in 1882. To make doubly sure that the legislature would not pass measures inimical to their enterprises, the great lumbermen hired lobbyists to guard and protect their interests; and occasionally men like Weyer-

[3] Stephenson, *Recollections,* 150, 190; *La Crosse Leader,* January 12, 1878; Kinney and Adler to Ingram and Kennedy, August 19, 1869, W. H. Day to Ingram and Kennedy, September 17, 1869, E. H. Playter to Ingram, Kennedy & Co., July 27, 1875, D. H. Cushing to Orrin H. Ingram, July 27, 1876, and Knapp, Stout & Co. to Ingram, July 14, 1885, in the Ingram Papers; Phelps to Holt and Balcom, April 12, 1887, in the Holt Lumber Company Papers.

haeuser visited the legislative halls in person to make known their wishes. Lobbyists were well supplied with money, and were disposed to spend it freely. "Send along a draft as finances melt easily," an Eau Claire lumberman was instructed by a lobbyist in 1878. When their common interests demanded it, lumbermen were able to pull together, putting aside the business competition that frequently placed them on opposite sides of the political fence. "I have taken the liberty," one lumberman-senator informed a friendly competitor in 1868, "to draw on you today for $150 to aid in paying the expense of lobbies here." Three years later the same senator led the legislative campaign for an act authorizing the construction of a dam on the Chippewa, a campaign that exploded into an airing of bribery charges by a joint committee of the legislature. Despite such setbacks the lumber lobby succeeded in obtaining many laws granting special privileges and franchises to lumbermen.[4]

With the same tactics they used to place friendly politicians in local and state offices, the lumbermen were usually able to send the right man to Congress too, but they discovered that control of the federal government was less easily achieved than control of the state. As a congressman wrote a lumberman constituent in 1885, "This is *not* so easy a place to get what one wants as is the Wisconsin legislature." Goaded by lumbermen to secure river improvements more rapidly, he exclaimed, "I do not *own* this Congress, and while I am willing to be held responsible for *Effort* no man can be responsible for *Results* in this mob."[5]

[4] Letters to Orrin H. Ingram from Alexander Meggett, October 2, 1867, C. R. Gleason, February 26, 1872, H. P. Graham, February 11, 1875, Thomas Carmichael, February 20, 23, 27, 1877, John C. Clarke, January 19, 1882, and William Miller, February 20, March 11, 1891, Joseph G. Thorp to Ingram and Kennedy, February 29, 1868, Thorp to Daniel Shaw, January 18, 1872, and Miller to C. A. Chamberlin, January 14, 1887, all in the Ingram Papers; Daniel Whitney to Moses M. Strong, December 12, 1852, in the Strong Lumber Papers; *Chippewa Herald,* February 10, 1872, November 9, 1877, December 22, 1899; *Eau Claire Free Press,* July 11, 1867; *Chippewa Falls Democrat,* October 28, 1869; Charles S. Bundy, *Early Days in the Chippewa Valley* (Menomonie, 1910), 6–9; Stephenson, *Recollections,* 193; Ise, *United States Forest Policy,* 371; *Wisconsin Senate Journal,* 1872, Appendix, 135–171, 178–181, 215–234.

[5] William Price to Orrin H. Ingram, February 6, 1885, and Price to Joseph G. Thorp and Ingram, February 20, 1885, in the Ingram Papers.

Many prominent lumbermen ran for congressman or senator themselves, confident that their training had prepared them for office. "The successful operation of the business of manufacturing or trading in lumber," wrote an Oshkosh newspaper editor in 1875, "requires and develops a wonderful amount of executive ability, and where men have been chosen from the ranks of our lumbermen to fill high positions of trust in the State or Nation we have yet to learn of a record marked with impotence or marred by dishonor."[6] Philetus Sawyer of Oshkosh, Isaac Stephenson of Marinette, Thaddeus C. Pound of Chippewa Falls, William Price of Black River Falls, and other lumbermen occupied seats in Congress, indifferent for the most part to the opposition's mutterings that it was "reasonable to suppose" they would devote their attention "largely to class legislation."[7]

The chief goals of the lumbermen in Congress and their friends at home were favorable land laws, a tariff on lumber, and river and harbor improvements. Of these, the last was their greatest concern because of the vital importance of transportation to the industry and the high cost of making the necessary improvements in driving and rafting streams and lake harbors. During an election campaign the congressional candidates in a lumbering district reiterated one promise oftener than any other: that they would secure larger funds for local improvements than their opponents. Sawyer was lauded as a great senator because he "managed to get through appropriations to make excellent harbors in Wisconsin where poor ones existed"; and Pound allowed his friends to proclaim that he was in Congress "to look after the many local wants of this great manufacturing district." Isaac Stephenson was not so naïve as he appeared when he wrote, "With indiscriminate denunciation of appropriations for rivers and harbors as 'pork-barrel' measures, I have no patience."[8]

[6] Oshkosh *Northwestern*, June 17, 1875.

[7] Oshkosh *Weekly Northwestern*, January 16, 1890; *Wisconsin State Journal* (Madison), January 10, 17, 1881.

[8] Stephenson, *Recollections*, 208; Oshkosh *Northwestern*, October 4, 1866; *Eau Claire Free Press*, August 31, 1882, January 21, April 15, 1886; *Chippewa Herald*, November 23, 1872, August 25, November 3, 1876, July 6, November 30, 1877, February 15, June 14, November 1, 1878, September 17, October 8, 1880, Janu-

Not only did the lumbermen-politicians work for the special legislation that would benefit their own business, but they supported other members of the business group who had the same interests. During the seventies and eighties they had been inclined to favor the regulation of railroads, but by the close of the century the two industries were pulling together. By that time rate discrimination against the lumber industry had been greatly reduced, and many lumbermen had made large investments in railway enterprises. Similarly they changed their attitude toward public finance. During the seventies, when the greenback was a major issue in national politics, the "saw-log statesmen," influenced by the Western debtor psychology in general and their own debtor position in particular, tried to strike a balance between hard and soft money. Pound voted to limit the issue of greenbacks to three hundred and fifty thousand dollars, and to make them exchangeable at par for gold—both anti-inflation devices—but he also favored their acceptance in payment of government bonds and customs duties, a favorite device of the inflationists. On the question of free silver he supported the Bland-Allison bill, a measure designed to satisfy the inflationists without actually producing much inflation. But by 1896 the lumbermen, now no longer in a debtor position, were opposed to any compromise on the currency issue. During the presidential canvass of that year many Wisconsin lumbermen, fearing the "appeals to class prejudice so constantly reiterated by Bryan and his friends," spoke of closing their mills or of raising the price of lumber if McKinley was not elected. As a group the lumbermen were solidly in favor of a gold standard.[9]

The narrowness of their political goals is evidence of the lumbermen's lack of any profound economic or political philosophy.

ary 13, August 24, 1888; *Report from Committee on Rivers and Harbors, Submitting H. 9824 to Make Regulations Governing Running of Logs and Rafts on Rivers* (56 Congress, 1 session, House Report no. 731, serial 4023, Washington, 1900).

[9] *Chippewa Herald,* November 16, December 7, 1877; July 5, 19, September 6, 27, 1878, December 29, 1882, August 28, 1896; *Eau Claire Free Press,* June 27, 1867, November 9, 1882; Oshkosh *Northwestern,* January 26, 1882; *La Crosse Leader,* June 18, 1870; La Crosse *Weekly Chronicle,* January 19, 1893; George B. Shanks to Orrin H. Ingram, August 18, 1893, and W. H. Day to Charles Horton, October 29, 1896, in the Ingram Papers.

Like most Americans of their class, they regarded government as a stimulator of business and the servant of the enterprising man. If they belonged to any school of political philosophy at all, it was that of Jeremy Bentham, the eighteenth-century Englishman who believed that government existed to provide the greatest good for the greatest number. This the government could best do by extending to businessmen like themselves such assistance as river improvements and protective tariffs, and otherwise keeping hands off. A bountiful Providence had supplied vast forests, large numbers of people who needed lumber, and men of prophetic vision to manufacture that lumber. As Joseph G. Thorp told the National Lumbermen's Association in 1876, "We have no right to waste this rich material . . . by adopting a policy of indifference, or by failing to use all the means available to so manufacture and use it as to accomplish the greatest good to the greatest number." He urged lumbermen to demand that the government provide better facilities for carrying on their business.[10]

Their conviction that Providence had assigned them a sublime role in history gave most lumbermen an almost smug confidence in themselves. Their industry was bound to become the wealthiest, their communities the most populous, in the country. The proceeds of a single day's log drive would "buy out an Eastern Town, Banks and all." Only men of "great capacity and breadth of view" could have made such splendid use of the opportunities presented by the frontier. Indeed, lumbering was the very keystone of economic stability in the region. As the Eau Claire lumberman Gilbert E. Porter put it in 1875:

> I need hardly assert that if enterprises other than the manufacture of lumber, have followed and are auxiliary to it, they have in the main, been made possible and profitable by it. The truth of this assertion is made perfectly clear by the known fact that every industry and profession is affected favorably or unfavorably in the same measure that manufacturing is prospered or depressed.[11]

[10] *Eau Claire Free Press,* September 28, 1876; Stephenson, *Recollections,* 163.

[11] *Eau Claire Free Press,* December 16, 1875 (from which the quotation is taken), September 28, 1876; *Chippewa Falls Democrat,* June 3, 1869, March 16, 1871; *La Crosse Independent Republican,* June 20, 1855; Stephenson, *Recollections,* 163.

Despite their utilitarianism, however, their self-confidence and pride of achievement, the lumber kings were not, for the most part at least, hedonists. They did not believe that the world existed solely to afford them wealth and pleasure. The worthiest of them manifested the highest sense of duty to their communities and high ethical principles in their business dealings. These they measured, of course, by the standards of the day; if some of them had trespassed on government land, it was not an offense for which their communities blamed them. Their neighbors were more likely to observe that they were conscientious churchgoers and that they met all their financial obligations promptly. More than that, they spent philanthropically at least part of the wealth that Providence and their own ability amassed for them. Philetus Sawyer and Orrin H. Ingram, for instance, gave substantial subsidies to Ripon College; Sawyer gave aid also to Lawrence College, as did Isaac Stephenson; Cadwallader C. Washburn and Stephenson endowed public libraries in their home towns.[12]

The fortunes made in lumbering or pineland speculation are more notable for their number than for their size. Of the thirty La Crosse men who in 1883 were worth over a hundred thousand dollars, seventeen were lumbermen; in the same year Eau Claire had fourteen lumbering fortunes of a hundred thousand or more. Among them was the millionaire Joseph G. Thorp, one of the founders of the Eau Claire Lumber Company, who maintained a large house in Eau Claire and another in Madison while he was a member of the state Senate. In the luxurious Madison home was a much-prized "parlor set" of Egyptian design, which was covered with purple satin and decorated with gilded and enameled sphinxes' heads. Hardly less magnificent was the grand piano that had been designed by Thorp's musical daughter. In this fabulously appointed house, which later became the state's executive mansion, the Thorps entertained the literary, musical, and political notables who visited Madi-

[12] Oshkosh *Northwestern,* June 17, 1875; E. R. Leland to Clarence Chamberlin, May 9, 1877, Isaac Stephenson to Orrin H. Ingram, February 8, 1898; Philetus Sawyer to Ingram, January 29, 1898, and Arthur Little to Clarence Chamberlin, March 11, 1877, in the Ingram Papers.

son. It was indeed the "center of Madison's gay life." One of the Thorps' greatest social triumphs was the marriage of their daughter to Ole Bull, the noted violinist. Revealing, too, is the *beau geste* Thorp made in 1870 when he gave his wife, on the eve of her departure for Europe, a well-publicized camel's-hair shawl which had cost five thousand dollars.[13]

Much less pretentious were the lives of the lumbermen's employees. Like the cowboy of the cattle country, the lumberjack has appealed to the novelist and romanticist, but when stripped of the glamor that has gathered around him, he emerges as a counterpart of the rough, uncouth pioneers that were found on every American frontier. Reveling in work that permitted him to display the splendid strength of his body, he also liked to throw himself into boisterous, energetic play. He loved rough horseplay, songs and stories, and garish costumes. Psychologically he combined the classic individualism of the frontier with a fine *esprit de corps* that enabled an expert woods crew to act as a unit. Mentally most of the lumberjacks were childish and naïve, but what they lacked in culture they made up in shrewdness, self-reliance, and stamina. Belonging to an age that had not yet become highly specialized, many of them were quite as capable of performing the work of blacksmith or carpenter as of cutting down a tree.[14]

In the camps the hardy lumberjacks followed their harrowing routine, living on the coarsest of fare, especially in the early days: pork and beans, bread, and tea. No fresh meat was available and few vegetables, since there were few farms in the neighborhood of the camps. In other lumbering sections of the country the men were sometimes afflicted with scurvy, but Stephenson could recall no outbreak of the disease in Wisconsin. As transportation facilities improved it became possible to extend the lumberjacks' diet to include vegetables and fruits, both dried and fresh, and butter.[15]

[13] *La Crosse Leader,* May 21, 1870; *Eau Claire Free Press,* August 2, 23, October 4, 1883, January 8, 1891, May 26, 1892; *American Lumbermen,* 1:321–322.

[14] Nelligan, *Life of a Lumberman,* 37–38, 107; Stephenson, *Recollections,* 81; *Report of the Bureau of Labor and Industrial Statistics,* 1885–86, p. 243; *Eau Claire Free Press,* February 14, 1884.

[15] Stephenson, *Recollections,* 85, 107; Otis W. Terpening to Charles E. Brown,

Isolated from civilization though they were, and often overworked, the lumberjacks nevertheless remained remarkably healthy. Other than the injuries caused by accidents, they had few ailments, and for these they developed their own simple remedies. Ten cents worth of epsom salts in a quart of whiskey "would cure about anything." Hinkley's Bitters took care of a common cold. When a man's toes became scaly from being wet too long, white lead was painted on. Rotted pine wood was used in lieu of talcum powder to soothe chafed skin. "Crumbs" (body lice) and "sidewheelers" could be eliminated either by mechanical means or by the use of blue vitriol. As time went on less primitive methods were introduced, and by the eighties many lumbermen were providing vaccination and other medical care for their camp crews.[16]

The ever difficult task of keeping order among the coltish lumberjacks was the responsibility of the camp foreman. Recalcitrants, who could not be turned out summarily into a snow-covered wilderness, often had to be convinced of the error of their ways with a strong arm and heavy fist. Some of the camps adopted a system of self-discipline under which a culprit convicted by his co-workers of breaking a camp regulation was fined a specified amount of tobacco and subjected to corporal punishment on the seat of his pants while he cowered on his hands and knees. Most lumbermen forbade the use of whiskey in camp, but the prohibition did not prevent fights among the men. Stories of brawls, riots, and even murders continued to trickle out of the woods.[17]

March 1, 1931, in the Brown Papers; Jones and others, *Lincoln, Oneida, and Vilas Counties*, 173. The supply list of a lumber camp in the Oconto district shows how varied the diet had become by the 1920's. It included canned peas, corn, peaches, milk, molasses, macaroni, vanilla, lemon extract, oatmeal, sugar, brown sugar, raisins, onions, dried peaches and apples, prunes, beans, cornmeal, canned tomatoes, graham flour, vinegar, lard, rice, cinnamon, cloves, allspice, mustard, butter, tea, and coffee. List dated February 9, 1927, clipped to payroll book of Holt Lumber Co., 1925–27, in the Holt Lumber Company Papers.

[16] Undated letter of Otis W. Terpening to Charles E. Brown, in the Brown Papers; *Eau Claire Free Press*, October 1, 1885.

[17] Stephenson, *Recollections*, 105, 130; Jones and others, *Lincoln, Oneida, and Vilas Counties*, 171; *Eau Claire Free Press*, March 1, November 29, 1883; Oshkosh *Northwestern*, February 1, 1861; Nelligan, *Life of a Lumberman*, 153.

Many responsible lumbermen realized that the primitive instincts of their isolated camp crews might be tamed by some sort of religious teaching, but found it impossible to arrange regular Sunday services in the woods. They did, however, lend their support to any organization that was willing to try. Some of them, like Orrin H. Ingram of Eau Claire, contributed funds to the Young Men's Christian Association, which in the 1880's sent traveling secretaries through the north woods to conduct religious meetings in the lumber camps. Wise in the ways of the lumberjack, these missionaries sought to make their meetings as attractive as possible before giving their brief talks and distributing the literature they had brought with them. The Women's Christian Temperance Union also sent emissaries into the woods, one of whom reported in 1893 that he had visited ninety camps in the course of the year and distributed two hundred free Bibles, thirty-five hundred "Christian papers," and three thousand leaflets. Many of the men took the W. C. T. U. pledge never to drink liquor again. What proportion of them kept their pledge is not a matter of record. In 1877 the lumberjacks of the Chippewa region were visited by a Seventh Day Adventist preacher who proclaimed that he was the prophet Elijah. Catholic nuns, trusting the reports that lumberjacks never molested a respectable woman, sometimes visited the camps to collect money for orphanages and other charitable institutions.[18]

The rude, hard life of the lumber camp naturally appealed only to men who had the physical equipment for rough work, long hours, and little rest. The lumberjack son of a Madison clergyman found it very discouraging. He must work from three o'clock in the morning, he wrote his brother, until seven or nine in the evening,

> and then after supper we roll into our soft, downy couch of lousy blankets, and lay and listen to the mocking bird, with music by the

[18] *Eau Claire Free Press,* January 4, 1877, January 4, September 20, 1883, March 3, 1887, April 6, 1893, May 2, 1895, February 16, 1899; Nelligan, *Life of a Lumberman,* 129; Knud Knudsen to Orrin H. Ingram, January 12, 1886, in the Ingram Papers.

entire band, and snoring in seven different languages, mostly imported—professional snores from Germany and Norway, warranted never to miss a note . . . while the beautiful odor of wet socks and foot rags is heard in the near distance, and finally fall to sleep to slow music, only to be awakened in a few minutes by the melodious voice of the cook, singing, "roll out your dead bodies, daylight in the swamp," etc. Then we get up and go to our beautiful and sumptuous repast of fricasseed pork and beans on the half shell with a basin of reduced ice water, flavored with copperas, and called by the low and uneducated, "tea." Such is life in the woods, but for me, give me six months, twice a year for two years, in Waupun, or some other place of enjoyment. Well, Ralph, I hope you will learn some useful trade of usefulness, that will keep you from ever having to go to the woods.[19]

The ordinary dress of the woodsman consisted of a short heavy coat, trousers of bright red, blue, or green mackinaw cloth, and, in winter, heavy woolen socks and rubbers or shoepacs which filled the bunkhouse with an indescribable redolence as they hung over the stove to dry in the evening. The raftsmen dressed even more ostentatiously. After the Civil War the "*costume de rigueur* of the raft pilot was French calf boots, black cassimere trousers, red flannel shirt of extra fine knitted goods, a large black silk necktie, tied in a square knot with flowing ends, and a soft, wide-brimmed black or white hat."[20] To this he might add when he was on holiday silver shirt studs the size of a half dollar, a resplendent sash with long fringes hanging on the right side, and a watch chain of heavy twisted links. Raftsmen liked the picturesque in names as well as dress, and such sobriquets as Buckskin Brown, Sailor Jack, and French Fred were common.[21]

What little time the lumberjack had for recreation in the woods—a few hours before bedtime and an occasional holiday, such as Christmas—he spent in playing rough games, telling stories, or chinning with his comrades in the bunkhouse. A

[19] *Chippewa Herald*, April 7, 1882.

[20] Durant, "Lumbering and Steamboating," in *Minnesota Historical Collections*, vol. 10, pt. 2, pp. 664–665.

[21] Bentley, Wisconsin River Lumber Rafting Days (MS.), 1–2; Robert K. Boyd, *Early Conditions of the Chippewa Valley* (n.p., 1921), 5.

WHi(D489)9631

Cable, Bayfield County, Wisconsin, probably about 1902–1908. Lumberjacks in the bunkhouse of Ole Emerson's lumber camp. Dry socks were very important for men working in the woods.

Whi(X3)142

Big Falls, Wisconsin, probably about 1890–1910. A cutting crew in the woods. Notice the tongs held by the man in the background; they were used to grab the end of a log so it could be pulled by the horse. Carr Collection.

typical game was one in which a blindfolded man stooped over and tried to guess who had struck him from the rear. The "greenhorn" was subjected to countless practical jokes. It was to get a laugh at the expense of a frightened young tenderfoot that veteran lumberjacks invented the now-famous mythical denizens of the pine woods: the Indian devil, the hodag, the high-behind, the satchel-jaw, and others. On special occasions energetic woods crews organized stag socials at which the usual country dances were performed by newly shaved lumberjacks, the pseudo-feminine character of whose two-hundred-pound partners was proclaimed by a grain sack tied around the waist or a handkerchief around the arm. For quieter oddments of time the ever-present deck of playing cards was pulled forth.[22]

It was during such recreational moments that the well-known tales of Paul Bunyan and his big blue ox were fabricated. These folk stories were not indigenous to the Great Lakes region, but were carried westward from New Brunswick and the Eastern states. Their appeal to the Wisconsin lumberjack can easily be exaggerated, but there can be no doubt that they were often retold and enlarged upon as time went on. Subjected to the frontier imagination, Paul became a giant of a man who could perform amazing feats of strength and dexterity. By tying a rope to the end of his axe he could cut forty acres of pine with a single swing, and it was thus that he cleared the Dakotas of trees. "If you don't believe me," the narrator would say, "go to the Dakotas. You won't find any pine." Babe, Paul's blue ox, was equally phenomenal. Measuring "forty-two ax-handles and a plug of tobbacco" between the eyes, it was capable of reversing the current of the Mississippi by taking a long cool drink. Indeed the Mississippi itself had been created by a leak in one of Paul's road sprinklers. "If you don't believe me, go West. The Mississippi River is there to prove it."[23]

Equally deserving of preservation as part of American folk

[22] A manuscript titled Forms of Amusement, and letter from Otis W. Terpening to Charles E. Brown, October 23, 1931, in the Brown Papers; *Eau Claire Free Press,* January 4, 1877.

[23] Otis W. Terpening, Myths of Paul Bunyan, a manuscript apparently written from memory by an old lumberjack, in the Brown Papers.

literature are the songs and ballads that the raftsmen of the great logging streams sang on their wearisome trips. In their free moments the rivermen loved nothing better than to jig to a lively tune or intone a sad ballad. Most of the river ballads radiated funereal melancholy, but a few shanty songs lilted with bright humor, and some were not for "ears polite." A favorite ballad was Billy Allen's *Shanty Boy*, a mournful dirge of a milliner's daughter who died of scarlet fever and a broken heart when her mother declared that "a shanty boy her daughter ne'er should wed." The handsome shanty boy himself

fell off a rapids piece,
On the falls at Mosinee
And ended thus his faithful love,
And all its miseries.
And now the old Wisconsin
Is rolling o'er his bones,
His companions are the Muskellonge,
His grave a bed of stones.

In the ballad of *Geary's Rock* the cruel waters of a rapids engulfed another brave raftsman. When they took him from his "watery grave,"

There was one fair form among them,
A girl from Saginaw town,
Her moans and cries did rend the skies
For her lover that was drowned.

A brighter picture was drawn of another legendary figure, a certain One-eyed Riley:

He was prime fav'rite out our way,
The women folks all loved him dearly.
He taught the parsons how to pray
An' got their tin, or pretty nearly.

There was also the amiable old Cap' Jones, who "jest took a drink . . . 'cause his innards was a-fyar," and the charming *Buffalo Gals:*

I'd like to make that gal my wife,
Gal my wife,
Gal my wife.
I'd be happy all my life
If I had her by me.

[Chorus]

Buffalo gals, can't you come out tonight?
Can't you come out tonight,
Can't you come out tonight;
Buffalo gals, can't you come out tonight,
And dance by the light ob de moon?[24]

Isolated for months from the rest of humanity, the lumberjacks developed a language of their own. Prunes became "logging berries," bread was "punk," and tea was "swamp water" or "belly wash." Working tools were also renamed, and a completely new vocabulary was invented to describe working operations. All this argot of the woods carried over into the home lives of the workmen. A lumberjack often spoke of his wife, sweetheart, or dog as a "prize log"; a man of low intelligence was a "dead head," the term used for a sunken log. To the outsider the language of the woodsmen was likely to be wholly incomprehensible, as may be judged from one lumberjack's description of an accident that had befallen him: "I was up in the sky with a crooked steel and I had a ground hop what was punk. Talk about sending up! He couldn't send up his best regards. We had a big schoolmam an' I told him to take a St. Croix, but he threw a sag into her and she gunned an' come back and broke out four of my slats." To the initiated his story meant that he had been top-loading logs with a cant hook, while his assistant remained on the ground and sent the logs up to him on a chain. One of the logs happened to be crotched, and the top-loader warned his helper to use a cant hook on the under side. Instead he placed it on top of the log so that the big stick slipped edgewise and broke four of the loader's ribs.[25]

[24] Bentley, Wisconsin River Lumber Rafting Days (MS.), 5; Copies of ballads collected by Charles E. Brown, in the Brown Papers; Russell, *A-Rafting on the Mississip'*, 207–213.

[25] Glossary of Slang in the Brown Papers; *Eau Claire Free Press*, July 12, 1877.

The sawmill town was much like any other small frontier town of the period: board walks, chipped and scarred by the calked boots of the rivermen; muddy streets flanked by rows of frame business houses and small and unpretentious dwellings; and an unduly large percentage of saloons and gambling houses unless the millowners in control of the town had forbidden the sale of whiskey, as they sometimes did. From beginnings like these some of the lumber towns grew into cities of size and importance. As early as 1861 La Crosse could boast two banks, six carriage and wagon factories, three cooperage shops, three foundries, and a sash and blind factory besides its sawmills. In 1864, when livestock was still running at large in its streets, Eau Claire had a sixty-thousand-dollar "opera house," streetcars, and three daily and five weekly newspapers.[26]

Dances, holiday celebrations, and charivaris were the principal social events in the lumbering towns. The ratio of men to women being about three to one, every woman for miles around was impressed into attending dancing parties and other social gatherings. In the 1870's log-rolling or "burling" contests began to be popular, and during the eighties baseball teams representing various mill crews supplied diversion for the town dwellers. Middle-class citizens found recreation and an outlet for their gregarious instincts in group activity of various kinds. A lodge that appealed strongly to Wisconsin lumber entrepreneurs during the 1890's was the Concatenated Order of Hoo-Hoos, whose membership was limited to lumbermen, newspapermen, railway officials, and persons engaged in the manufacture or sale of woodworking machinery. Its professed object was to promote the health, happiness, and long life of its members. Its emblem was a black cat; the initiation fee was $9.99; and its annual meeting, which lasted nine hours and nine minutes, began on the ninth day of the ninth month at nine minutes past nine o'clock.[27]

[26] Nelligan, *Life of a Lumberman,* 37; *La Crosse Weekly Democrat,* December 20, 1861; *Eau Claire Free Press,* October 30, 1884, October 26, 1899.

[27] Stephenson, *Recollections,* 148–149; *Eau Claire Free Press,* August 13, 1874, July 12, 1883, October 26, 1893, March 15, 1894, September 6, 1900; La Crosse *Republican and Leader,* March 2, 1872, August 23, 1873; La Crosse *Weekly Chronicle,* August 19, 1886.

At the close of the work season lumberjacks, drivers, and raftsmen, surfeited with labor, their season's pay burning holes in their pockets, swarmed into these towns—as many as a thousand perhaps overrunning a village of fifteen hundred inhabitants. Eager to sample every pleasure offered, many of them, shaved and newly outfitted, entered upon a drunken debauch that lasted until their money was gone, while sober citizens prayerfully awaited their departure. Saloons and gambling halls, some of which were prepared to cater to every desire of the work-weary men, did a thriving business. To the accompaniment of brassy music, vaudeville performances were repeated ad infinitum, lunch counters and bars were kept busy, and gaming tables were thronged. And amid the boisterous crowds ubiquitous dancing girls plied their trade.[28]

The lumberjacks themselves usually sought only innocent, if violent pleasure; it was said that no woodsman, however intoxicated, would deliberately insult a respectable woman or steal money. But besides the lumberjacks themselves, definitely shady characters were attracted by the presence of easy money during the spending season. Fights, shootings, and stabbings occurred frequently, and in many places it was unsafe to be on the streets after dark. Accounts of murder and robbery committed by these unsavory hangers-on of saloons and gambling halls frequently enlivened the pages of the local weeklies. The police force of a community often found itself helpless to handle the situation when a drunken brawl developed into a riot. In any event it was felt, even by honest townsmen, that the woodsmen should not be curbed too much in their quest for a good time, lest another year they take their money elsewhere.[29]

The logging frontier was of course typical of the frontier in

[28] James Ginsell to Ingram and Kennedy, June 17, 1868, in the Ingram Papers; Russell, *A-Rafting on the Mississip'*, 4, 60, 189, 212; Oshkosh *Weekly Northwestern*, May 8, 1890; *Chippewa Herald*, February 2, 1894, January 11, 1895, January 19, March 23, 1900; *Eau Claire Free Press*, March 24, 1898, March 22, April 5, 1900; Nelligan, *Life of a Lumberman*, 59–60; Lincoln, Lumbering in Early Days (MS.), 4–5.

[29] *La Crosse Weekly Democrat*, November 22, 1861; *Chippewa Herald*, March 29, 1895, March 18, 1898, October 27, 1899; *Eau Claire Free Press*, June 19, 1873; Nelligan, *Life of a Lumberman*, 96–97, 182–188.

general, where violence of every sort was common. The mysteriously burned sawmills, log stealing, boom cutting, blown-up dams, and lynchings were only the specific manifestations of the lawlessness with which every pioneer community had to cope for a time. Alteration and forgery of log marks plagued the lumbermen just as brand-changing plagued the ranchers of the cattle country; looting followed in the wake of every disastrous forest fire; and what would now be termed the hijacking of rafts and even steamboats bedeviled many law-enforcement officers. In time, however, as on the other frontiers, these conditions gradually came to an end. Married woodsmen, who used their wages to establish themselves in farming, were one of the stabilizing influences. The population of the logging regions could not have been fundamentally vicious, for such men would not have accepted the hard work required of the lumberjack. Contemporary opinion attributed to bad whiskey, probably with ample justification, most of the crimes that disturbed the peace of the lumbering towns.[30]

To the men and women who lived and worked in the pine-scented atmosphere of the industry, lumbering often became more than a business or a job. It was a subtle, hypnotic essence, seeping from forest and mill into the very fabric of their existence, sometimes gently alluring, sometimes cruelly dominating, but always persuasive and often seductive. Few have expressed this attitude better than the doggerel artist who envisioned just what he would do when he reached heaven:

I'll stand by the gate and keep watching for those
Who come with the smell of the pine on their clo's.
For even in heaven I'll want it, I will,
The smell of the sawdust that comes from the mill.[31]

[30] S. J. Carpenter to Moses M. Strong, April 23, 1861, in the Strong Lumber Papers; Augustus Cole to Holt and Balcom, May 27, 1887, in the Holt Lumber Company Papers; *Chippewa Falls Democrat,* August 5, 1869, May 11, 1871; *Chippewa Herald,* May 31, 1873, April 24, June 19, 26, 1874, May 14, 1880; *Eau Claire Free Press,* July 9, 1868, April 8, July 29, 1875, November 24, 1881; Oshkosh *Northwestern,* May 30, 1872, July 22, 29, 1875; Nelligan, *Life of a Lumberman,* 38–41, 44–45, 106–107, 290; Bruncken, *Forests and Forestry,* 84–85.

[31] Douglas Malloch, "Smell of the Sawdust," in *La Crosse County Historical Sketches,* Series 3: *The Lumber Industry,* 57.

14. An End and a Beginning

WITH every passing year after the cession of the Indian lands until almost the close of the nineteenth century the production of sawed lumber increased in Wisconsin except during periods of economic depression. From an estimated cut of somewhat less than two hundred million board feet in 1853 the annual output mounted to about a billion and a quarter feet in 1873 and to a maximum of well over four billion feet in 1892. After that the inevitable decline began. In 1897 a skilled forester estimated that there remained in Wisconsin only fifteen billion board feet of standing white pine and enough Norway pine and hemlock to bring the total of merchantable coniferous timber to twenty-nine billion board feet; in addition about sixteen billion feet of hardwoods were available. This was scarcely enough to maintain the industry for a decade at the current rate of production.[1]

Many discerning experts had begun to realize soon after 1870 that the rapid destruction of Wisconsin's pine forests by axe and fire would before long deprive the state of its leadership in production. Nearly every year some prophet predicted the imminent disappearance of the state's pine resources. In 1873 the press published an estimate circulating in Pennsylvania that the supply in Wisconsin could last only thirty years longer. A year

[1] Hotchkiss, *Lumber and Forest Industry*, 733; Usher, *Wisconsin*, 1:157.

later Isaac Staples, prominent Stillwater lumberman, prophesied that lumbering would come to an end in twenty-five years. In 1883 an Eau Claire editor foresaw the extinction of white pine lumbering in only seven years. When these dire predictions failed to materialize, many lumbermen became unduly optimistic about the future. In 1897, despite the contrary opinion of a forester, Chippewa valley lumbermen were expecting to cut logs on their river and its tributaries for at least another twenty years, although the steady reduction in the annual cut forced them to concede that the end was in sight. In 1900 Wisconsin was still leading all states in the production of white pine lumber, but the state commissioner of industrial statistics observed that most lumbering counties had seen their years of maximum cut. The production of white pine lumber had declined fifteen per cent in the preceding decade, a decrease which the commissioner attributed to the dwindling of good standing timber. Disastrous forest fires aided the axemen in bringing the timber supply of Wisconsin close to exhaustion. Thus by 1906 the lumber industry was entering its final stage.[2]

Long before white pine lumbering actually ceased, changes in the industry and in the state's economic life had begun to reflect the imminence of this exhaustion. Lower grades of timber began to be used, the ownership of timberland became more concentrated, and the larger companies began a policy of conservation designed to protract their business as long as possible. To the veteran logger the most notable change was the increasing harvest of the once-despised hardwoods, hemlock, and cedar. Hardwood logging began to assume sizeable proportions during the early seventies, and by 1876 hardwood and hemlock lands were booming. In the early nineties some mills that had been cutting only pine announced that henceforth they would cut

[2] *Chippewa Herald,* March 1, 1873, July 3, 1874, January 27, 1882, August 14, 1885, May 28, 1897, September 1, October 27, 1899; Warder, "Forests and Forestry," in *Transactions of the Wisconsin State Horticultural Society,* 1880–81, p. 153; *Eau Claire Free Press,* November 8, 1883, April 17, 1890, February 2, 1899; Filibert Roth, *Forestry Conditions and Interests of Wisconsin* (Forestry Division, United States Department of Agriculture, Bulletin no. 16, Washington, 1898), 78; *Report of the Bureau of Labor and Industrial Statistics,* 1903–04, pp. 310–311, 321, 337; *Report of the State Forester,* 1906, p. 6.

nothing but hardwoods. By 1899 the hardwood and hemlock cut was slightly over nine hundred million board feet, and by 1906 it had increased to more than a billion board feet. It still continues, but since 1920 has been consistently less than a billion feet. The last sizeable stands of white pine in Wisconsin were cut during the 1930's.[3]

As they were forced to reduce their business in Wisconsin, many lumber barons began buying standing timber in the South or on the Pacific coast. Investments in Southern yellow pine were being made as early as the 1870's, and by the nineties they were considerable. The Illinois Central cooperated by conducting tours of capitalists through the Southern pine region, and men like Weyerhaeuser and his associates bought thousands of acres there. By 1885 Douglas fir and redwood on the west coast had also begun to attract Wisconsin lumbermen. Tours conducted by the Pacific coast railroads led to the purchase of millions of feet of standing timber by Wisconsin syndicates. Before 1900 many Wisconsin lumbermen were cutting and sawing timber on the coast, maintaining their old businesses in Wisconsin at the same time. The mining projects and power developments in the far West also interested lumbermen who had surplus capital to invest.[4]

With the decline of white pine lumbering in Wisconsin many communities once dependent on it began to promote agriculture. In 1895 the state legislature lent its assistance by creating

[3] *Report of the Bureau of Labor and Industrial Statistics*, 1903–04, pp. 338–339; *Wisconsin Lumberman*, 1:256 (January, 1874); La Crosse *Republican and Leader*, March 21, 1874; Oshkosh *Northwestern*, February 20, 1879; *Chippewa Herald*, December 22, 1882; La Crosse *Weekly Chronicle*, March 19, 1885; *Eau Claire Free Press*, May 13, 1886, May 15, 1890, December 8, 1892; H. G. Chichester to Orrin H. Ingram, August 15, 1892, and M. H. McCarthy to Ingram, August 16, 1898, in the Ingram Papers; W. A. Holt to Holt and Balcom, February 23, 1887, in the Holt Lumber Company Papers; [Wisconsin] Committee on Land Use and Forestry, *Forest Land Use in Wisconsin*, 38.

[4] *Chippewa Herald*, April 22, 1877, August 11, 1882, August 10, 1885, September 28, 1888, February 1, April 26, 1889, December 14, 1894; letters to Orrin H. Ingram from H. C. Davis, February 21, 1893, James J. Hill, February 25, 1893, George B. Burch, October 18, 1894, and W. L. Ewart, August 2, 1899; W. H. Day to D. M. Dulaney, Jr., March 21, 1896, and W. H. Day, Jr., to Col. W. Thomas Hart, January 6, 1898, all in the Ingram Papers; *Eau Claire Free Press*, August 6, 1891, April 20, July 13, October 26, 1893, November 26, 1896, November 9, 1899, June 21, 1900; *La Crosse Tribune and Leader-Press*, April 17, 1938.

a board of immigration to encourage the settlement of northern Wisconsin, and several lumbering counties set up their own boards of immigration and published "handbooks" for the prospective settler. Such agencies as the Wisconsin Immigration and Development Association, organized and largely supported by lumber and land companies, attracted thousands of immigrants to the state, as did the sale of tax lands by the northern counties. Collectively these efforts amounted to an informal campaign, which resulted in the establishment of many new agricultural communities. It proved that the sandy white pine and jack pine lands could not be successfully cultivated, but areas which had once borne mixed hardwood and coniferous trees were often fertile enough, and the new agriculture became a permanent economic asset of the state.[5]

As the lumber industry declined, the population of the chief logging counties also ceased to grow. Cities and villages that had been wholly dependent on lumbering now faced inevitable decay or complete obliteration. The population of Ashland County, which had increased almost twelve hundred per cent in the decade of the eighties, increased less than nine per cent from 1890 to 1900, and actually declined fourteen per cent between 1920 and 1930. The population of Bayfield County, which likewise had increased almost twelve hundred per cent in the decade after 1880, had a growth of less than eight per cent from 1910 to 1920, and declined during the twenties. Eau Claire County, the population of which had trebled in the sixties and almost doubled in the seventies, increased only about three per cent in the nineties. There were "dozens of cities and villages," observed the state Forestry Commission in 1898, "where the inhabitants have begun to wonder what will become of them when the

[5] *Eau Claire Free Press,* October 15, 1896; *Chippewa Herald,* January 4, 1895, November 23, 1900; Schafer, *Agriculture in Wisconsin,* 140–142; *Hand Book of Chippewa County, Wisconsin,* prepared by the [State] Board of Immigration (Milwaukee, 1896), 4; Soo Line, *Wisconsin and Its Opportunities* [1911], 7; *Wisconsin Immigration and Development Association: Its Purposes, Endorsements and What the Milwaukee Press Says of It* (Milwaukee, 1906); *Annual Report to Members of the Wisconsin Immigration and Development Association for Year Ending September 11, 1907* (Milwaukee, 1907).

timber is gone and the mills close down. Everybody has seen settlements very prosperous ten years ago, which are now abandoned by almost all their former inhabitants."[6]

Not all sawmilling centers declined, however. Many of them managed to survive as retail and distributing centers for the surrounding farming community, often with diversified industries. The lumber industry itself, in response to diminishing and new types of raw materials, underwent an evolution in the direction of greater refinement of product. Declining supplies meant progressively lower net profits, and lumbermen had no choice but to close their plants or modify their product. Those who were reluctant to abandon their Wisconsin businesses because they owned valuable rights, homes, and other property at the scene of their operations gradually turned to the manufacture of more highly finished products and the use of wood other than pine.[7]

Some cities and villages, by virtue of their location and previous industrial development, had an advantage over others in the struggle for continued existence. As early as 1855, hardly five years after the founding of the village, La Crosse was clamoring for a pail factory, plow factories, a machine shop, and many other industrial establishments. At the same time plans were being made for developing the town as a wholesale center. Throughout the period when lumbering was at its height, efforts were made to introduce other industries. During the decade of the nineties, when the value of the lumber output at La Crosse declined more than fifty per cent, the products of other industries increased thirteen per cent. In 1900 thirteen were reported to be flourishing in the town. Ten of these had increased their production over that of 1890: agricultural implements; furniture; confectionery; foundry and machine-shop products; brewing; publishing; baking; plumbing, gas, and

[6] *Report of the Forestry Commission of Wisconsin*, 1897–98, pp. 15–16; [Wisconsin] Committee on Land Use and Forestry, *Forest Land Use in Wisconsin*, 36; *Eau Claire Free Press*, May 1, 1890, November 26, 1891, March 23, 1899, January 18, June 7, November 1, 1900; *Chippewa Herald*, November 24, 1893.

[7] Whitbeck, "Industries of Wisconsin," in *Annals of the Association of American Geographers*, 2 (1912):61–62.

steam fitting; tobacco, cigars, and cigarettes; and planing-mill products.[8]

In the region around Lake Winnebago crude lumbering declined after 1870, and the sawmilling centers began to develop other manufactures. Oshkosh, a typical lumbering town, soon became a center for the production of sashes, doors, and blinds; matches; wagons and carriages; trunks and other luggage; and furniture. All these products were wood derivatives, but by 1880 industries not dependent upon wood began to be established, among them meat packing, flour milling, iron products, and corsets. During the next decade Oshkosh was the country's largest producer not only of sashes, doors, blinds, and matches, but also of wagons and carriages. Between 1890 and 1900, when the manufacture of lumber declined thirty-seven per cent in Oshkosh, total manufactures increased about two per cent, and several products increased by as much as thirty per cent: carriages and wagons, foundry and machine-shop products, and furniture. By 1900 Oshkosh was producing more than a third of all the cigars and cigarettes manufactured in Wisconsin.[9]

The city of Eau Claire carried on during the eighties a vigorous campaign to attract new industries to the town. By 1892 it had, in addition to its nine sawmills, a pulp and paper mill, a furniture factory, a trunk factory, three sash, door, and blind plants, seven planing mills, a barrel and stave factory, two breweries, two book binderies, a sawworks, and three carriage and wagon factories. Superior had in 1900, when its sawmill product was still increasing, malt liquor, plumbing, gas and steam-fitting, publishing, and foundry and machine-shop enterprises. By 1910 there were in the state twenty-one separate industries which utilized wood as their chief raw material. Most of these were represented in northern Wisconsin, whose industrial life had by now been almost entirely reshaped. Many of the new factories were direct descendants of earlier sawmills, just as several of

[8] *La Crosse Independent Republican,* June 4, August 8, 15, 1855; La Crosse *Republican and Leader,* April 25, 1874; *Report of the Bureau of Labor and Industrial Statistics,* 1903–04, pp. 359–360.

[9] Schafer, *Winnebago-Horicon Basin,* 268, 278, 279, 290; *Report of the Bureau of Labor and Industrial Statistics,* 1903–04, pp. 367–368.

the new automobile factories were the outgrowth of vehicle factories.[10]

Had logging and sawing been carried on more efficiently, Wisconsin's white pine industry could undoubtedly have been prolonged. It is estimated that a billion more board feet could have been produced in the years 1872–1905 had band saws been used to the exclusion of muley and circular saws. Besides what was lost because of needlessly wide kerfs, much valuable material was burned as refuse. Logging crews were responsible for more spoilage; they sometimes left high stumps in the woods, and much timber was wasted because of the fashion in which the trunk was cut into logs. Before 1880 hemlock that had been cut for its bark was left to rot in the woods, when it might have been converted into lumber and pulpwood. Windfalls and partly decayed trees that could have been sawed into lumber were allowed to remain in the forests and furnish fuel for forest fires. Young growth was often cut for making roads and for other purposes, and much young timber was destroyed by falling trees.[11]

Fire, too, levied a tremendous toll on Wisconsin's forests, largely because loggers usually took only the choicest parts of the trees, leaving the rest to dry into excellent kindling. If all slash had been piled and carefully burned, it is conceivable that Wisconsin might still have a great lumber industry. As it was, perhaps more good pine timber was burned than ever reached the sawmills. Forest fires broke out almost every year, being especially dangerous after a long drought. The Peshtigo fire of 1871, blaz-

[10] *Eau Claire Free Press,* January 26, 1882, August 19, 1886, May 6, 1892; Whitbeck, "Industries of Wisconsin," in *Annals of the Association of American Geographers,* 2:62; *Report of the Bureau of Labor and Industrial Statistics,* 1903–04, pp. 315–316, 371. The articles of manufacture made chiefly of wood included agricultural implements, beekeepers' supplies, boats, boxes, chairs, church and school furnishings, cigar boxes, coffins and caskets, excelsior, furniture, tool handles, musical instruments, refrigerators, pulp, sashes, doors and blinds, toys, lawn furniture, trunks, office fixtures and cabinet work, vehicles and vehicle stock, windmills, tanks and silos, woodenware, and various minor products. Franklin H. Smith, *A Study of the Wisconsin Wood-Using Industries* (Madison, 1910), 10, 12, 17–44.

[11] Dopp, "Geographical Influences in the Development of Wisconsin," in the *Bulletin of the American Geographical Society,* 45:745–746.

ing up after a severe dry spell, devastated in four hours an area ten miles wide and forty miles long. More than a thousand persons, trapped by the rapidly advancing holocaust, lost their lives. Observers remarked that the air was so dry as to seem almost inflammable itself. In 1885 nearly all of an eighty-mile strip in the Wisconsin River valley was swept by fire. In 1905 fires broke out in nearly every one of the thirty-two counties of northern Wisconsin, a total of 1,435 in all, which burned over more than a million acres and sent up a pall of smoke that sometimes impeded navigation on Lake Michigan as far south as Chicago. Sixty per cent of them had been set by farmers burning brush or trying to clear their pastures of stumps, fifteen per cent by sparks from locomotives, and twenty-five per cent by hunters, campers, logging crews, and others passing through the woods.[12]

As Wisconsin's forests gradually disappeared and citizens became more generally aware of the waste caused by lumbermen and by fires, a demand arose for the conservation of the remaining forest resources. In the sixties and seventies the movement was sponsored chiefly by intellectuals and fanatics who incurred only the contempt of practical lumbermen and of such politicians as United States Senator Timothy Howe, who once remarked that the "generation yet unborn" had done nothing for him and he did not care to "sacrifice too much" for them. Congress and the state legislature had demonstrated for many years, in their land policies, that they were indifferent to conservation ideas.[13] But the conservationists refused to be discouraged.

They could cite a long series of reports and recommendations which proved that not all government officials had been indifferent to the problem. As early as 1866 Commissioner Wilson of the General Land Office had warned that the amount of standing timber in the Great Lakes area was "so diminishing as to be

[12] *Ibid.*, 746–747; Hotchkiss, *Lumber and Forest Industry*, 296–297; *Chippewa Herald*, October 12, 19, 26, December 19, 1871, August 6, 1875, May 10, 1889, October 20, 1893; *Eau Claire Free Press*, January 18, 1872, May 1, 1879, October 8, 1885, June 3, 1886, September 28, 1893; La Crosse *Republican and Leader*, May 16, 1874; La Crosse *Chronicle*, August 19, 1886; Hough, *Report on Forestry*, 3:231–239; *Report of the State Forester*, 1907–08, pp. 58–62; 1909–10, pp. 114–117.

[13] Ise, *United States Forest Policy*, 46–53; Bruncken, *Forests and Forestry*, 96–98.

WHi(X3)12327

Angelica, Wisconsin, 1875. Upham Brothers sawmill built by William and Charles Upham after the Peshtigo fire of 1871 to salvage the burnt-over timber of the area. Note the burned timber in the background of this picture. Photo by A. Scidmore, Appleton, Wisconsin.

WHi(X3)44753

Vicinity of Lost Lake, Sawyer County, Wisconsin, about 1895. A rutter at work; this machine was pulled through the woods to make grooves in which sled runners would fit, thus making heavily loaded log sleds easier to pull and control.

a matter of serious concern." A year later the legislature of Wisconsin provided for the appointment of three commissioners to report on "facts and opinions" concerning the effects of such clearing of the land and "the duty of the state in regard to the matter." The commissioners, headed by the able and respected Increase Lapham, accepted the false notion of the day regarding the evil effects of forest depletion on climate, but their report contained sound warnings. Citing Egypt and Palestine as examples of the horrible results produced by the destruction of all the trees, they asserted that such destruction would raise the temperature of the ground in summer and lower it in winter, that it would make the ground drier, diminish the flow of water from springs, increase the "suddenness and magnitude of floods," cause soil erosion, impair the productiveness of the soil, and increase the violence of thunderstorms. With Benthamite practicality the Lapham commission urged that, in the interest of the "greatest good to the greatest numbers," the state extend its authority into the field of forest management. If the state could regulate hunting seasons, it certainly could regulate the time and place of the cutting of its trees. The commissioners recommended also that the state experiment with tree culture and plant a series of tree belts. In 1868 the legislature passed a general timber-culture act, offering a bounty for the successful cultivation of timber, but this ended for several years the state's efforts to do something about preserving its forests.[14]

In Congress conservation measures received short shrift during the seventies, eighties, and nineties. As in the state government, lumbermen and other interests opposed to any restriction of forest exploitation were securely entrenched. Public opinion, however, slowly came to the aid of the conservationists. In 1875 the American Association for the Advancement of Science memorialized Congress on the rapid depletion of the nation's forest resources. Wisconsin newspapers, even those in the heart of the lumber regions, began questioning the wisdom of destroying the forests without making some provision for the future. In 1874

[14] *Report of the Commissioner of the General Land Office,* 1866, p. 33; Lapham and others, *Destruction of Forest Trees,* 9–36; Ise, *United States Forest Policy,* 97–98.

Governor Cadwallader C. Washburn, one of whose many business interests was lumbering, warned the owners of pineland who were cutting large quantities of timber each year that they were killing the goose that laid the golden egg. Articles on European forestry systems began to appear in newspapers and magazines. Farmers in northern Wisconsin began to realize that their taxes would be heavier when the lumbermen were forced to depart for lack of timber. In 1884 the Carriage Builders' National Association, fearing the eventual disappearance of their chief raw material, petitioned Congress to encourage tree planting and to discourage the wanton destruction of the existing forests. Ever stronger became the demand for the withdrawal of the national forest lands from the market. In 1893 a speaker told the Wisconsin Horticultural Society that the forest policies of the state were "the methods of barbarism." The Wisconsin State Forestry Association, organized at the same meeting to encourage conservation, immediately began a vigorous campaign for the conversion of the remaining state lands into forest reserves. In 1895 Governor William Upham stated in his message to the legislature that the demands of the conservationists were attracting state and national attention. Most important of all, many Wisconsin lumbermen had by 1899 come to favor state forestry legislation.[15]

Under the pressure of public opinion and the reports of successive secretaries of the interior and commissioners of the General Land Office, the attitude of Congress toward conservation slowly changed also. The passage of the Timber Culture Act, inadequate as it was, is proof that there was at least some conservation sentiment in Congress as early as 1872. A law of 1876

[15] *Eau Claire Free Press,* May 15, 1873, June 17, 1875, April 17, 1879, October 12, 1882, December 27, 1883, October 1, 1889, January 23, 30, 1890, April 13, November 9, 1893, November 29, 1894; La Crosse *Republican and Leader,* August 15, 1874, March 23, 1878; Oshkosh *Northwestern,* August 16, 1883, February 7, 1884; *Chippewa Herald,* September 13, 1889, December 14, 1894, January 28, 1898; Carriage Builders' National Association, *Memorial to the Senate and House of Representatives of the United States* (New Haven, 1884), 4; J. J. Blaisdell, *Forest and Tree Culture in Wisconsin,* an address delivered before the Wisconsin Horticultural Society at Madison on April 6, 1893 (n.p., n.d.), 17; *Governor's Message,* 1895, p. 13; William J. Starr to Orrin H. Ingram, February 8, 1899, in the Ingram Papers; Ise, *United States Forest Policy,* 39, 371.

required the commissioner of agriculture to appoint an expert to study the production and use of lumber and other forest products. His appointee was Franklin B. Hough of Lowville, New York, a pioneer student of forestry, who submitted a voluminous report stressing the need for conservation measures. Three years later a special committee was appointed by authority of Congress to classify the public lands, codify the land laws, and make recommendations for a future policy. In the preparation of its report the commission visited almost every Western state and undoubtedly was responsible for much of the legislation that followed. During the eighties two concrete plans for the conservation of forest resources on a national scale were presented to Congress by government officials. Both provided for the withdrawal of the public timberlands from the market and their transfer to the jurisdiction of the forestry bureau that had been established under the commissioner of agriculture. Finally, in 1891, Congress carried out a general revision of all the land laws and empowered the president to set aside forest reservations in any state or territory comprising public lands wholly or partly covered with timber. The fight had not yet been won, but the foundation had been laid for a rational and beneficial forestry policy.[16]

As the nineteenth century drew to an end the state of Wisconsin developed a forest policy of its own. By 1896 fire and trespassers had destroyed all but twenty per cent of the fifty thousand acres of good timberland which had been set aside in 1878 for a state park. An early law providing a penalty for the inten-

[16] *Special Report on the Subject of Forestry by the Commissioner of Agriculture* (45 Congress, 2 session, House Executive Document no. 24, serial 1802, Washington, 1877); Edward A. Bowers, *Report Submitting a Plan for the Management and Disposition of the Public Timber Lands* (50 Congress, 1 session, House Executive Document no. 242, serial 2560, Washington, 1887); *Public Timber Lands* (50 Congress, 1 session, House Executive Document no. 144, serial 2558, Washington, 1888); *Message from the President of the United States Transmitting Report Relative to the Preservation of the Forests on the Public Domain* (51 Congress, 1 session, Senate Executive Document no. 36, serial 2682, Washington, 1890); Ise, *United States Forest Policy*, 38, 43–44, 92–98, 109–115; Hough, *Report upon Forestry*, [vol. 1] (Washington, 1878; not included in the serial set of congressional documents); vol. 2 (46 Congress, 2 session, House Executive Document no. 37, serial 1922, Washington, 1880); [vol. 3] (47 Congress, 1 session, House Miscellaneous Document no. 38, serial 2046, Washington, 1882).

tional or careless setting of a forest fire had proved to be quite ineffectual. In 1895, with a growing awareness of conservation sentiment, the legislature passed an act creating a system of town fire wardens to prevent and fight forest fires. Two years later the legislature ordered the remainder of the state park to be sold, and set up a new forestry commission to study methods of classifying the state lands and of withdrawing from the market all lands better suited to forests than to agriculture. This commission, consisting of George B. Burrows, H. C. Putnam, and Ernest Bruncken, interviewed many lumbermen in the northern part of the state and in 1898 recommended the legal regulation of timber cutting, experimental tree culture, and the creation of a department of forestry to administer a system of state forests.[17]

The legislature did not act upon these recommendations until 1903, when it created an unsalaried forestry commission, provided for the appointment of a superintendent of forests, and set aside a forest reserve of sixty-two thousand acres in Forest, Oneida, Vilas, and Iron counties. The legislature's intention was to set aside, as the nucleus of a forest reserve, what remained of the state lands elsewhere, but it neglected to repeal the laws governing their sale, and the new commission was helpless to prevent the disposal of unreserved forest land. In 1905 another law was passed which substituted a nonpolitical forestry board for the commission and set aside as a forest reserve all state lands, except school lands, north of townships numbered thirty-three. Of this land the board might sell any that was unsuitable for forestry purposes and use the proceeds to buy timbered areas. The next year the state acquired from the federal government, in conformity with a congressional act introduced by Senator Robert M. La Follette, twenty thousand acres of vacant federal land for forestry purposes. Wisconsin now had a reserve of 254,072 acres.[18]

[17] *Report of the Commissioners of School and University Lands,* 1880, p. 23; *Eau Claire Free Press,* June 3, 24, September 30, November 18, 1897; *Chippewa Herald,* February 15, March 22, May 31, 1895, March 12, May 28, 1897; *Revised Statutes,* 1878, ch. 72, sec. 4406; *Report of the Forestry Commission of Wisconsin,* 1897–98.

[18] *Report of the State Forester,* 1906, pp. 5, 8, 10–15, 32; *Session Laws,* 1903, ch. 450; 1905, ch. 264.

The state presently took steps to strengthen its forest policy at several points. In 1907 ten thousand dollars was appropriated, the first of a series of such appropriations, to enlarge the forest reserve by the purchase of county-held tax lands, and a year later the forestry board was authorized to purchase other than tax-title lands. To encourage private owners to plant their lands in forest trees the legislature offered tax exemption for thirty years on any amount up to forty acres. Finally, the Wisconsin Valley Improvement Company was authorized to build dams and storage reservoirs along portions of the Wisconsin River for the purpose of achieving a more uniform stream flow. Since this project was to be carried out in conjunction with the creation of state forests at the headwaters of the stream, the state forestry board was given partial control of the company. In 1908 a special conservation commission recommended an increase of the state's forest reserves and its purchase of the cutover lands held by lumber companies.[19]

A study made by the state forester in 1909 showed that from forty-five to ninety-five per cent of the area of each of the ten most northerly counties consisted of cutover land and that much of the rest had been culled for pine. Some of the larger lumber companies were neither selling their cutover land nor holding it for a second crop. The state forester concluded that they were presumably retaining title in order to perpetuate their control of the town government and thus insure that taxes on their still-timbered property would be kept at a minimum. The forester had already expressed the view that the state was taxing its timber to death and that irrational inconsistencies of assessment existed; some lands were taxed at sixteen per cent of their true value, others at fifty per cent. In some areas every forty was assessed at twelve hundred dollars, regardless of its true value. Moreover, taxes were increasing every year and so forcing the owner of timberland to cut his trees rapidly rather than conserve them, a policy to which the fear of destructive fires contributed. In one northern county a single forty on which $5.51 had been

[19] *Report of the State Forester*, 1907–08, pp. 21–24, 92–93, 100–102; *Session Laws*, 1907, ch. 96, 335, 592.

paid in taxes in 1904, was taxed $44.50 only four years later. Such drastic increases were of course partly the result of a rapid advance in stumpage values, but also of the fact that communities in the forested areas were growing and were demanding more improvements. Many citizens of these sections, foreseeing that the depletion of the forests would deprive them of their best source of revenue, were demanding that the largest possible sums be raised immediately, not only to cover current expenditures but to finance improvements the need for which was merely anticipated. A sudden rise in the tax rate was often due to the fact that the lumber companies had lost control of local politics.[20]

Uncertainty respecting the amount of future taxes, the risk of damage by fire and insects, and fluctuations in the price of lumber tended to discourage the private owner from making the long-term investment required for the reforestation of his lands. Studies of the growth rate of coniferous trees indicated that nearly fifty years of development would be required to produce a tree that could be cut. No capitalist could afford to wait that long to realize a return on his investment. The lumber manufacturer did, however, support the cause of conservation by practising economy in his cutting and sawing operations. The old wasteful methods of logging had largely been abandoned by 1905, and by 1910 many manufacturers had learned to utilize their waste products to the best advantage. In that year at least one manufacturer was making use of defective hardwoods and of the tops of trees down to an inch and a half limb. Treated in a chemical plant, one cord of this material produced sixty bushels of charcoal, ten gallons of wood alcohol, and one hundred and forty-five pounds of acetate of lime. To demonstrate effective utilization of such mill waste and to study allied problems, the United States Forest Products Laboratory—the only institution of its kind in the country—was established at Madison in 1909 through the cooperation of the state university with the federal forest service.[21]

[20] *Report of the State Forester,* 1909–10, pp. 37–39, 41–45; Alfred K. Chittenden and Harry Irion, *The Taxation of Forest Lands in Wisconsin* (Madison, 1911).

[21] *Report of the State Forester,* 1909–10, pp. 49–50; Woodward, *Valuation of*

A scientific forestry program such as that upon which Wisconsin had embarked takes time to develop. But notable results have been achieved. Whereas in 1931 less than a third of the thirteen million acres in the state useful chiefly for forests could be considered "competent forest lands," today a well-planned program of tree planting and fire protection is in operation. Eight state forests have been established—most of them since 1924—which comprise a quarter of a million acres of forest land. Another two million acres are held by twenty-seven counties as county forests. Wisconsin is no longer in danger of losing all its growing timber.[22]

* * * * *

During the years of its dominance and financial strength the lumber industry affected almost every aspect of the state's economy. It built towns and cities, roads, bridges, and railways. It cleared lands that could later be used for agriculture. It gave employment to settlers and supplied a market for farmers, thus helping to populate the state. It brought money as well as people from other states. It contributed a substantial share of the revenues required for the administration of state and local government—in 1890, for example, a sum equal to the entire tax collection of the state government. It fostered the establishment of many secondary industries using wood as a raw material, especially in communities that had once been sawmilling centers. And it conferred immense benefits upon the treeless plains west of the Mississippi as they were invaded by westward-moving throngs of migrants.

But the industry also left a less salutary heritage. Its ruthless attack on the forests reduced large tracts to virtual deserts. When the forests were gone, the state discovered that the flow of its streams had been disturbed and that in some sections the soil was

American Timberlands, 35; Smith, *Wisconsin Wood-Using Industries,* 7; Land Economic Inventory Division, Wisconsin Department of Agriculture and Markets, *Growth Studies of the Three Principal Species of Conifers Found in Vilas County, Wisconsin* (mimeographed, [Madison, 1930]), 2; Dopp, "Geographical Influences in the Development of Wisconsin," in the *Bulletin of the American Geographical Society,* 45:749.

[22] Wisconsin Chamber of Commerce, *Referendum No. 2 on the Report of the Committee on Conservation on a State Land Policy* (Milwaukee, 1931); W. A. Ti-

being washed away. Presently, too, it was facing a lumber famine as a result of the selfish and unscientific policies of earlier administrations; as early as 1910 it was being forced to import a large part of the lumber it needed. Viewed in the light of such destruction, the lumber kings, protected by the state and federal governments, would seem to have reaped profits out of all proportion to the benefits they conferred. Contrasted with the opulence and power they enjoyed, largely as a result of the "unearned increment" in the value of their lands, the hardship and poverty borne by the average logging jobber, lumberjack, and mill hand presents a dismal picture. Even so, due credit must be given to the successful lumberman who, through his own energy, stamina, and courage, raised himself from obscurity to a position of wealth and power. He faced risks that would have daunted the average man; and if he rode the tides of prosperity on to fortune, he did so with the full knowledge that financial shoals lay all along the course. The most depressing result of the lumber industry, and perhaps the most needless of all, was the plight of the settler on submarginal cutover lands. Cajoled by promoters into investing his savings in land suited only to the growing of trees, he carried on a hopeless struggle to eke out a living by farming. A soil survey of the state forest reserve in 1910 indicated that only about a fifth of the land could be profitably tilled. It is to be fervently hoped that the remaining four-fifths will some day revert to the trees for which it is so eminently fitted.[23]

tus, "Two Decades of Wisconsin Forestry, 1905–1925," in the *Wisconsin Magazine of History,* 30:187–191 (December, 1946); *The Wisconsin Blue Book,* 1950, pp. 272–274. In addition to the publicly owned forests, 160,000 acres of timber or unimproved land are still a part of the trust funds set aside by the constitution for purposes of public education; these are managed by the commissioners of public lands. *Ibid.,* 222.

[23] *Report of the State Forester,* 1909–10, p. 56; [Wisconsin] Committee on Land Use and Forestry, *Forest Land Use in Wisconsin,* 37; Hotchkiss, *Lumber and Forest Industry,* 735.

Bibliography

Bibliography

THE materials for the study of the lumber industry, which are both extensive and varied, include much scattered data not readily available for reference. The list below includes not only those that are most accessible and most important for the study of the industry in Wisconsin but also most of the others used in the preparation of this study. Excluded are a few printed materials which have been cited in the footnotes only for the information they furnish on some specific point.

NEWSPAPERS

For virtually every aspect of the Wisconsin lumber industry the newspapers of the principal lumbering towns furnish a running account of activities in the region, editorials revealing local sentiment on questions affecting the industry, and occasional historical sketches. The papers of Eau Claire and Chippewa Falls, probably because lumbering dominated the economy of these towns more than that of other communities of comparable size, give the best coverage. The *Eau Claire Free Press* for the years 1867–1900 and the *Chippewa Herald* of Chippewa Falls for approximately the same period provide information not only on local conditions and events but also on lumbering developments throughout the state and nation. The *Herald* is especially valuable for a study of Weyerhaeuser and his associates.

These two journals have been used as the basic newspaper sources, but other papers throughout the state have supplied more detailed information on developments outside the Chippewa valley. Almost without exception the papers of the lumbering communities were politically biased and were slanted more or less strongly in favor of sawmill owners, or of some particular group of millowners, but for the most part they can be relied upon for purely factual information. In addition to the two papers mentioned, the most useful are the following:

Chippewa Falls Democrat, 1869–1872
La Crosse *Chronicle,* 1880–1894
La Crosse Democrat, 1861–1866
La Crosse Independent Republican, 1854–1859
La Crosse *Republican and Leader,* 1871–1879
Northwestern Lumberman (Chicago), 1876–1879
Oshkosh *Northwestern,* 1865–1893
Wisconsin Lumberman (Milwaukee), 1873–1875

MANUSCRIPTS

The best sources of information on the internal problems of the industry and on the attitudes of lumbermen toward the issues of their day are unpublished manuscripts in the possession of the State Historical Society of Wisconsin. Of these the Ingram Papers constitute the most important single collection, by virtue of the span of years they cover and the large number of letters from other firms that are included among them. Covering the period from 1857 to 1904, they comprise a record of the long business career of Orrin H. Ingram of Eau Claire. Naturally most of the letters deal with Chippewa valley and upper Mississippi River business, but the incoming letters contain much valuable information on lumbering affairs throughout the state. The outgoing correspondence, in the form of letterbook press copies, is informative on the business practices of the firm. Unfortunately it has been greatly damaged by water. The collection is almost unique in the amount of data it contains on the political activities of lumbermen.

Much political information may also be gleaned from the Strong Lumber Company Papers, covering the years 1838–90, but the collection is more valuable as a source of data on business problems. Moses M. Strong of Mineral Point was a politician, promoter, and speculator who wrote garrulously of his experiences. The Lumber Papers constitute only a portion of this collection; another portion, labeled General Correspondence, 1838–1841, is helpful on the land problem.

For the study of business practices the Holt Lumber Company Papers, covering more than a century, 1839–1943, are extremely valuable. The collection is an extensive one, including not only incoming and outgoing correspondence but also ledgers, journals, cashbooks, records of purchases, inventories, and miscellaneous records of other kinds. Despite the lacunae that exist in the collection, it furnishes data on accounting practices, profits and losses, land costs, and lumber prices over a long period. The Holt Lumber Company, originally Holt and Balcom, operated logging camps in the Oconto valley, a sawmill at Oconto, and a wholesale yard at Chicago. Much of the correspondence in the letterbooks passed between the company's Oconto and Chicago offices, and, though partially water-spoiled, it gives as does no other collection an insight into the day-to-day problems of running a lumber business.

Other useful manuscripts, including reminiscences, are the following:

BABBITT, T. T. Comparison of the Lumber Industry of Wisconsin with That of the [Pacific] Northwest. Eight-page typewritten narrative, dated 1921, by a veteran lumberman who had moved from Wisconsin to Oregon. Largely a description of logging practices.

BARTLETT, WILLIAM W. Copies of Letters Written from Camps of the Park Falls Lumber Company and the Hines Lumber Company in 1921. Devoted largely to descriptions of the logger's life.

BENTLEY, STEPHEN R. An Account of Wisconsin River Lumber Rafting Days. A fourteen-page manuscript written in 1911 by a former lumberman of Sauk City, Wisconsin.

BRADLEY AND RICHARDSON. Journal and ledger kept by this firm of jobbers, showing numerous entries against Carson and Eaton on the Eau Galle River, 1845–49.

Brown Papers. Miscellaneous manuscripts collected over a number of years by Charles E. Brown, former curator of the museum of the State Historical Society of Wisconsin. Contains valuable material on the social history of lumbering.

CARSON AND EATON. A ledger kept by this lumbering concern on the Eau Galle River, 1846–51.

Eau Claire Lumber Company Papers, 1846–1888. A small collection, but includes excellent examples of various types of contracts, patents, deeds, and receipts.

Executive Records: Lumber Inspector, 1861–1905. Letters sent to the governor's office by and concerning state lumber inspectors.

Executive Records: Timber Agents' Reports, 1858–1890. Data on trespass and the attitude of the public on the practice.

Hamilton Papers, 1855–1890. Alfred Kittredge Hamilton was engaged in lumbering at Fond du Lac until 1883. A small collection.

session, House Miscellaneous Document no. 38, serial 2046, Washington, 1882. Statistics on forestry conditions and on the manufacture of forest products, with strong arguments for conservation.

Loose Logs, Steamboats, and Rafts on Certain Rivers and Streams. 56 Congress, 1 session, Senate Report no. 1035, serial 3890. Washington, 1900. A short document outlining the conflicting interests of steamboat owners and log drivers.

Lumber Rafts on Great Lakes. 51 Congress, 2 session, House Report no. 3759, serial 2888. Washington, 1891. Brings out some of the difficulties involved in lake transportation.

Owen, David Dale. *Report of a Geological Reconnoissance of the Chippewa Land District of Wisconsin.* 30 Congress, 1 session, Senate Executive Document no. 57, serial 509. Washington, 1848. Interesting narrative and description as well as scientific report.

Report on Manufacturing Industries in the United States at the Eleventh Census: 1890. 52 Congress, 1 session, House Miscellaneous Document no. 340, pt. 22, serial 3026. Washington, 1895. Census figures on Wisconsin lumbering before 1900 are of doubtful validity, but this is an excellent survey made near the height of the Wisconsin industry.

Roth, Filibert. *Forestry Conditions and Interests of Wisconsin.* Division of Forestry, United States Department of Agriculture, Bulletin no. 16. Washington, 1898. This careful survey by a skilled forester was also published, under the title *On the Forestry Conditions of Northern Wisconsin,* by the Wisconsin Geological and Natural History Survey (Bulletin no. 1, Economic Series no. 1, Madison, 1898).

Sources of the Mississippi, Saint Croix, Chippewa, and Wisconsin Rivers. 45 Congress, 3 session, House Executive Document no. 54, serial 1858. Washington, 1879. An Army Engineers survey containing material of value to the historian of lumbering.

Tunell, George G. *Statistics of Lake Commerce.* 55 Congress, 2 session, House Document no. 277, serial 3679. Washington, 1898. This is a work of much broader scope than the title suggests. It constitutes what amounts to a brief history of trade and navigation on the Great Lakes.

Waukegan Harbor and Oconto and Wolf Rivers. 46 Congress, 2 session, House Executive Document no. 19, serial 1913. Washington, 1881. Another Army Engineers survey containing incidental data of value.

STATE DOCUMENTS

The *Journal* of the Assembly and that of the Senate, published annually until 1883 and biennially thereafter, provide the outline of the legislative history of bills and resolutions, but except for an occasional appendix reporting a committee investigation they yield little or no data on the content of proposed legislation or on debate.

Before 1873 the session laws were published annually in two parts: *General Laws* and *Private and Local Laws.* Beginning in that year they were combined in *Laws of Wisconsin,* published annually until 1883 and biennially thereafter. Several collections and indexes facilitate the use of these series:

Compilation of United States Laws, State Laws, Decisions, Messages, and Correspondence Relating to the Swamp and Overflowed Land Fund of the State of Wisconsin. [Compiled by Ernest G. Timme, secretary of state.] Madison, 1882.

HAGENAH, WILLIAM J. *The Law of Labor in Wisconsin.* Madison, 1908.

Legislative and Congressional Acts Concerning the Land Grants by the United States to the State of Wisconsin. U. S. Circuit Court for Western District of Wisconsin, Madison [1878].

A Synoptical Index of the General and Private and Local Laws of Wisconsin from the Organization of the Territory to 1873 Inclusive. Madison, 1873.

Wisconsin Lien Laws upon Real and Personal Property, Logs, Lumber, Timber, Bark, Boats, and Vessels, with Notes and Forms. Madison, 1878.

Revised Statutes of the State of Wisconsin were published in 1849, 1858, 1871, and 1878, and supplements in 1852 and 1883. The revised statutes of 1898 bear the title *Wisconsin Statutes.* These compilations provide not only convenient summarizations of law on various subjects connected with lumbering but also references to earlier legislation in the volumes of session laws. Interpretations of the law by the Supreme Court of Wisconsin have been published since 1848 in the *Wisconsin Reports,* where the cases are indexed alphabetically by name.

The publications of various state departments and agencies also contain much data on lumbering, but are not of equal value. The most useful are the following:

401–412, 490–499, 585–609, 653–663, 736–749, 831–846, 902–920 (June–December, 1913). Applies to Wisconsin some theories of Friedrich Ratzel and Ellen C. Semple.

DURANT, EDWARD W. "Lumbering and Steamboating on the St. Croix River." *Minnesota Historical Collections,* vol. 10, pt. 2, pp. 645–675. St. Paul, 1905. Informative and reliable account by a former steamboat pilot.

DURRIE, DANIEL S. "The Public Domain." *History of La Crosse County, Wisconsin,* 210–230. Western Historical Company, Chicago, 1881. A summary of legislation pertaining to the public lands of Wisconsin.

ELLIS, ALBERT G. "Upper Wisconsin Country." *Wisconsin Historical Collections,* 3:435–452. Madison, 1857. A rousing contemporary account of early logging and rafting activities on the Wisconsin River. Ellis was editor of the *Wisconsin Pinery,* a weekly published at Stevens Point.

FERNOW, BERNHARD E. "American Lumber." In Chauncey M. Depew, ed., *1795–1895: One Hundred Years of American Commerce,* vol. 1, pp. 196–203. New York, 1895. An excellent survey by a chief forester of the United States.

FISH, CARL RUSSELL. "Phases of Economic History of Wisconsin, 1860–1870." *Proceedings of the State Historical Society of Wisconsin,* 1907, pp. 204–216. A general picture of labor and industry during the Civil War decade.

FOLSOM, WILLIAM H. C. *Fifty Years in the Northwest.* St. Paul, 1888. A memoir covering many aspects of pioneer life, including lumbering. Reliable for the general picture but not always for precise dates and other details.

GATES, PAUL W. *The Wisconsin Pine Lands of Cornell University: A Study in Land Policy and Absentee Ownership.* Ithaca, New York, 1943. The best scholarly work on the problem of land ownership in Wisconsin.

GLOVER, WILBUR H. "Lumber Rafting on the Wisconsin River." *Wisconsin Magazine of History,* 25:155–177, 308–324 (December, 1941; March, 1942). The best study of river rafting in Wisconsin.

GREGORY, JOHN. *Industrial Resources of Wisconsin.* Chicago, 1853. Promotional literature.

GREGORY, JOHN G., ED. *West Central Wisconsin: A History.* 4 vols., Indianapolis, 1933. Most useful for its generous quotations from firsthand accounts.

Henry, William A. *Settlers' Handbook of Northern Wisconsin.* Madison, 1895. By the dean of the College of Agriculture of the University of Wisconsin. Contains a useful soil map.

Hibbard, Benjamin H. *A History of Public Land Policies.* New York, 1924. A reliable reference work for all aspects of land policy.

Holley, John M. "Waterways and Lumber Interests of Western Wisconsin." *Proceedings of the State Historical Society of Wisconsin,* 1906, pp. 208–215. Brief sketch of logging on the Wisconsin, Black, Chippewa, and St. Croix rivers before the Civil War.

Hotchkiss, George W. *History of the Lumber and Forest Industry of the Northwest.* Chicago, 1898. By a trade journalist of many years' experience who collected much of his material by personal contacts with lumbermen.

Ingram, Orrin H. *Letters of a Pioneer.* Eau Claire, 1916. A slight, but useful, supplement to the Ingram Papers.

Ise, John. *The United States Forest Policy.* New Haven, 1920. Scholarly and reliable.

Jones, George O., and others, eds. *History of Lincoln, Oneida, and Vilas Counties, Wisconsin.* Minneapolis, 1924. Better than most commercial county histories.

Keyes, Willard. "A Journal of Life in Wisconsin One Hundred Years Ago." *Wisconsin Magazine of History,* 3:339–363, 443–465 (March, June, 1920). Keyes, a native of Vermont, came West in 1817 and spent his second winter logging at Black River Falls. His account of that experience begins on page 453.

La Crosse County Historical Sketches, series 3: *The Lumber Industry* [edited by Albert H. Sanford]. La Crosse, 1937. Contains a series of articles by different authors on various aspects of lumbering.

Larson, Agnes M. *History of the White Pine Industry in Minnesota.* Minneapolis, 1949. A detailed and dependable account by a scholar. Contains much material on the Wisconsin industry.

Lockwood, James H. "Early Times and Events in Wisconsin." *Wisconsin Historical Collections,* 2:98–196. Madison, 1856. See pages 132–141 for an account of logging and milling in Indian territory about 1830.

Martin, Lawrence. "Progressive Development of Resources in the Lake Superior Region." *Bulletin of the American Geographical*

Index